# Evolution

## A modern synthesis

**W H Dowdeswell,** M.A., F.I.Biol.

Emeritus Professor of Education, University of Bath

Heinemann Educational Books Ltd
22 Bedford Square, London WC1B

LONDON   EDINBURGH   MELBO        D
HONG KONG   SINGAPORE   KUA           NEW DELHI
IBADAN   NAIROBI   JOHANNESBURG
PORTSMOUTH (NH)   KINGSTON   PORT OF SPAIN

First published 1984

ISBN 0 435 60227 6

Cover and text illustration by Sam Denley
Typeset by Eta Services (Typesetters) Ltd, Beccles, Suffolk
Printed in Great Britain by Fletcher and Son Ltd, Norwich

# Contents

## The author and publisher would like to acknowledge the following sources from which tables were taken and figures were redrawn with permission:

Table 1.2, Tribe, M. A. *et al* (1981), *The Evolution of Eukaryotic Cells, Studies in Biology No. 131*, Edward Arnold; Fig. 2.5, Patterson, C. (1978), *Evolution*, Routledge and Kegan Paul PLC; Fig. 2.10, Simpson, G. G. *et al* (1959), *Life*, Routledge and Kegan Paul PLC; Fig. 2.12, Williamson, P. G., in edn by Maynard Smith, J. (1982), *Evolution Now*, Nature Macmillan; Figs. 4.6 and 4.11, King, R. C. (1962), *Genetics*, Oxford University Press; Figs. 6.2 and 6.4, Wilson, E. O. *et al* (1973), *Life on Earth*, Sinauer Associates Inc.; Table 6.1, Tivy, J. (2nd edn, 1982), *Biogeography*, Longman; Fig. 6.8, Patterson, C., in edn by Maynard Smith, J. (1982); *Evolution Now*, Nature Macmillan; Fig. 6.19, Patterson, C. (1978), *Evolution*, Routledge and Kegan Paul PLC; Fig. 6.20, Gribbin, J. and Cherfas, J. (1982), *The Monkey Puzzle*, Bodley Head; Fig. 7.1, Darlington, C. D. (1953), *The Facts of Life*, Allen & Unwin; Fig. 8.2, Kerkut, G. A., *The Invertebrata*, Cambridge University Press; Fig. 8.3, Young, J. Z. (2nd edn, 1962), *The Life of Vertebrates*, Oxford University Press.


## Acknowledgement is also made to the following for permission to reproduce photographs:

The National Coal Board (Fig. 2.1); Camera Press Ltd. (Fig. 3.1); Natural History Photographic Agency (Figs. 3.3, 3.4, 7.2); Gene Cox (Fig. 5.4); Hulton Picture Library (Fig. 4.1); Frank Lane (Fig. 6.7); Jacana (Figs. 2.2, 2.6, 2.8, 2.9, 2.11); Dr M. A. Tribe (Fig. 1.6); S. Beaufoy (Fig. 5.8); Professor A. D. Bradshaw (Fig. 5.15); Dr J. Thorpe (Fig. 1.8); Julie Moss (Fig. 4.8).

# Preface

A hundred years after the death of Charles Darwin, evolutionary biology is once again a focal point of controversy and debate among scientists. Some, perhaps the majority, would maintain that neo-Darwinian theory is sufficient to accommodate all the fresh evidence that has come to light during the past twenty years or so. Others adopt a more radical stance, suggesting that major adjustments of viewpoint are now needed.

*Evolution – a Modern Synthesis* aims to provide a synthesis and evaluation of the existing situation. Its coverage includes present views on the origin of life, the geological record and the problems it poses, Darwinism and its modern derivations, and evolution above the species level (including that of man) with particular reference to problems of taxonomy. Considerable space is devoted to a discussion of alternative explanations of change both scientific (such as Lamarckism) and religious (including fundamentalism and the creationist movement). It is hoped that the book may provide stimulating reading for students in higher education at all levels, sixth formers, and others with a basic knowledge of biology who may wish to study evolution further.

For those unfamiliar with the subject or whose background is somewhat rusty, a summary is included at the end of each chapter, and also a specific list of references and topics for discussion. The topics are intended either as individual self-checks or for debate in groups. At the end of the book is a glossary, which defines most of the technical terms used (these are also explained in the text), and also a select bibliography of more general works on evolution.

This book is the outcome of many years of teaching evolution at all levels, combined with extensive research in the field of ecological genetics. Frequent discussions with colleagues and students have contributed materially to its contents and the way in which they are presented. In this connection, I would like particularly to thank John Barker, Tessa Carrick, and Professor Arthur Lucas for the trouble they have taken in reading the typescript and for many valuable criticisms and suggestions.

It is a pleasure to thank Tony Wheeler of Bath University for the excellent enlargements he has made from my negatives, and also Elaine Cromwell for the patience and efficiency with which she has prepared the typescript. My thanks are also due to all those who have permitted me to reproduce data, tables, and diagrams from their published work.

W. H. Dowdeswell
1984

# 1

# Life and non-life

Precisely how the Earth came into existence is still a matter for speculation. However, there is much less doubt about when the event took place; modern estimates now agree that the Earth's age is about 4600 million years.

When studying the possible origins of life from non-life, we may well ask how measurements of such a staggering magnitude can be made. During the past three centuries, our concepts of the duration of the Earth's existence and methods of determining it have greatly changed. In the early eighteenth century Archbishop Ussher estimated the period from the Creation to the birth of Christ based on the chronology of the Bible and arrived at a figure of 4004 years. By the beginning of the nineteenth century, the great geologists James Hutton and Charles Lyell had opened up a new dimension in space and time through their studies of the processes of geological change taking place simultaneously over the Earth's surface – such as sedimentation, erosion, and fluctuations in land and sea levels. As a result, it soon became clear that the Earth must be many millions of years older than had previously been supposed. As we shall see in Chapter 3, these discoveries greatly assisted Darwin in formulating his ideas on the origin of species.

Radioactive isotopes and their decay products have provided scientists today with powerful and much more precise methods of dating than were available to the early geologists. One of the best known and most widely applied methods is carbon-14 dating, the principles of which are as follows. High-energy cosmic-ray particles (mainly protons) are constantly bombarding the Earth's atmosphere from outer space. A nitrogen-14 atom absorbing one of these high-velocity particles is converted to carbon-14. Living organisms represent a vast pool of carbon, most of it in the form of carbon-12 but some as carbon-14. The ratio of carbon-12 to carbon-14 ($^{12}C/^{14}C$) in life is assumed to be the same as that of the atmosphere because of the exchange that takes place through the carbon cycle. But in death, alteration in the ratio occurs due to radioactive decay and because there is no longer any metabolic exchange with the surrounding environment. The total carbon pool of the Earth, taking into account living organisms, carbon dioxide (as $CO_2$ gas and in solution in water), and carbonate ($CO_3^{2-}$) in minerals, has been estimated at $1.2 \times 10^{-12}$ g $^{14}C$ g$^{-1}$ total carbon, consisting of all the isotopes of carbon present. The half-life of carbon-14 is $5.7 \times 10^3$ years (5730). To date a particular sample, its $^{14}C$ content is compared with that of the total carbon pool. The difference represents the level of decay and hence its age. Carbon-14 dating is, however, limited because breakdown is relatively rapid and the amounts that remain soon become too depleted for precise estimation. In practice its limit is about 10 half-lives or about 57 300 years.

Estimation of the age of the oldest igneous rocks necessitates the use of isotopes with a slow rate of decay, some of the most widely used being those of uranium ($^{238}U$ and $^{235}U$) and thorium-232. Their breakdown products are the various isotopes of lead (see Table 1.1). In rocks containing appreciable quantities of uranium or thorium, the ratio of these elements to the amount of lead present will

1

**Table 1.1** Lead isotopes in rocks containing uranium and thorium

| Parent substance | Half-life (years) | Lead isotope |
|---|---|---|
| None known | — | Lead-204 |
| Uranium-238 | $4.5 \times 10^9$ | Lead-206 |
| Uranium-235 | $7.13 \times 10^8$ | Lead-207 |
| Thorium-232 | $1.39 \times 10^{10}$ | Lead-208 |

provide an estimate of age. The older the rock, the lower will be the uranium/thorium content and the higher the lead. Alternatively, since uranium-235 has a shorter half-life than uranium-238 (see Table 1.1), the ratio of $^{207}Pb/^{206}Pb$ can be used as a geological clock. The assumption here, as with all other radioactive dating, is that the elements used were, indeed, primeval components of the Earth's surface or atmosphere.

Several lines of evidence have confirmed the conclusions from the uranium–lead method that the Earth's age is about 4600 million years. Thus, analysis of meteorites – offshoots of our Solar System and therefore the same age as the Earth – gives a similar result. Again, the use of other slowly decaying isotopes, such as potassium-40, which changes to calcium-40 and argon-40, have substantiated the findings of the uranium–lead technique.

## The first organic compounds

The element carbon is an essential component of all living things. One of the key happenings in the sequence of events that led to the origin of life must therefore have been the formation of simple organic compounds. In order to speculate on how these may have arisen we must turn once again to the evidence provided by ancient rocks. One of the most striking things about them is that although they contain a wide variety of different metals, metallic oxides are not present. Oxygen gas, which is highly reactive, must therefore have been absent from the original Earth's atmosphere, and we have good evidence that it remained so until the advent of photosynthesis by plants at a much later date.

Geological deposits indicate that in its early years, the Earth's surface was partly covered by areas of shallow, fresh water. These must have been murky with volcanic ash, which covered much of the Earth. The average surface temperature was probably around 25 °C (77 °F). There is much evidence of volcanic activity at this time and it is likely that electric discharges caused by thunderstorms were also frequent. Since no oxygen was present, the layer of ozone that now surrounds the Earth and absorbs much of the ultraviolet solar radiation will have been absent. There was therefore no shortage of energy of all kinds. The early atmosphere was composed largely of nitrogen, carbon monoxide, hydrogen, and water vapour, but in the presence of so much free energy it is unlikely that these would have remained unchanged for long. Thus a typical reaction for nitrogen could have been:

$$N_2 + 6H_2 \overset{energy}{\rightleftharpoons} 2NH_3$$

Like all chemical reactions, this one is reversible, the balance of products depending upon the conditions of the reactants. Thus, at about room temperature

(25 °C), provided hydrogen was plentiful, most of the nitrogen will have been converted into ammonia. This would quickly have been dissolved in any water present to form ammonium hydroxide ($NH_4OH$), giving rise to strongly basic conditions. But, as Folsome [1.1] has pointed out, the early igneous rocks contained considerable quantities of silicates, typical examples being the various kinds of quartz ($SiO_2$) and these could not have precipitated in such an alkaline environment. Problems such as this serve to highlight the danger of taking too simple a view of conditions in the early atmosphere and of treating each likely chemical reaction as a separate and isolated entity. In fact, the opposite situation must have existed with all the atmospheric components mixed together and interacting with one another in different ways. The fundamental question is whether such an enormous diversity of chemical events occurring simultaneously could have produced the kind of organic compounds needed to provide a basis for the origin of life.

One way of testing such a hypothesis is to attempt to simulate, at least in part, the kind of conditions thought to prevail when the Earth was formed. In 1953, Miller and Urey [1.2] working at the University of Chicago, devised a straightforward set of experiments based on the apparatus shown in Fig. 1.1. This consisted of a gas chamber fitted with electrodes into which were introduced mixtures of such gases as ammonia, methane, and hydrogen. Water vapour was also supplied from the flask at the bottom left of the apparatus. A spark discharge was passed between the electrodes at 60 000 volts for several days, an energy equivalent to that of some 50 million years on the original Earth. The resulting products were analysed and found to contain a range of organic molecules including some of potential biological importance, such as the amino acids glycine, alanine, and glutamic acid.

It could be argued that these early experiments only involved one form of energy, and a limited range of starting materials. Later investigations greatly

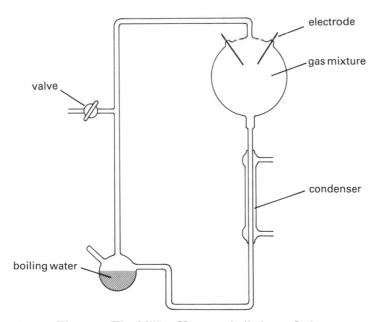

**Fig. 1.1** The Miller–Urey spark discharge flask

expanded this approach using different proportions of reactant gases and other sources of energy, particularly ultraviolet light. The overall results can be summarised as follows:

  (i) In the presence of high quantities of energy, gases such as carbon monoxide, nitrogen, hydrogen, and water vapour can form reactive intermediate compounds such as ammonia and methane.
 (ii) In similar conditions, these intermediates can react further to form a wide variety of organic molecules.
(iii) Among these organic products are at least thirteen amino acids of biological importance in the formation of proteins, also the nitrogen bases adenine, thymine, cytosine, guanine, and uracil, which are the essential constituents of deoxyribonucleic acid (DNA) and ribonucleic acid (RNA) – basic components of the cell nucleus. To what extent these compounds, once formed, could have avoided subsequent decomposition in the conditions of the early Earth is, of course, impossible to ascertain.

As was emphasised earlier, the complexity of the situation that must have existed in primeval times precludes the possibility of simulation in a single experiment. But the outcomes of many different investigations have together brought us much nearer to an understanding of the first steps that took place towards the origin of life.


# Origin of the first cells

There is a huge gap between the production of individual organic molecules such as amino acids with potential biological significance and their assemblage into the integrated system which we regard as the basic unit of life – the cell. In order to bridge this gap there are two essential requirements:

  (i) The concentration and assembly of appropriate molecules in an orderly manner.
 (ii) The development of a method of self-replication – the dominant feature which distinguishes life from non-life.

Regarding the first of these, several explanations have been put forward to account for the concentration and aggregation of molecules from the dilute 'soup' that must have existed for millions of years at the Earth's beginning. One of the most plausible of these is based on the presence of polymers; large molecules formed by the joining together of two or more similar molecules. These are frequently produced on the surface of water after spark discharge experiments such as those of Miller and Urey [1.2], their number increasing exponentially with time, one structure serving as a focus for the assembly of others. So-called **protocells** could thus have been found in primeval ponds alongside the organic molecules of which they were composed. They were, however, far removed from cells as we know them.

An important deficiency of protocells is any mechanism for the transport of genetic information and hence, by implication, for replication. Following the discoveries of Watson and Crick in 1953 [1.3] we now know that the genetic apparatus occurring universally among living organisms (with the exception of viruses containing RNA) is the molecule of deoxyribonucleic acid (DNA). This consists of two closely related parts:

(i) A double-stranded polymer of pentose (five-carbon) sugar and phosphate units arranged in two long chains wound round one another spirally like two coiled springs (Fig. 1.2). A single chain may be more than 10 000 units long.

(ii) Joining the two chains, a series of cross-connections that link pairs of organic bases together, rather like the rungs on a ladder (Fig. 1.2(a)).

A single unit of base, sugar and phosphate is known as a **nucleotide**. Along the length of the DNA chain, the nucleotides are so distributed as to achieve symmetry on the two sides of the 'ladder'. Thus, the bases adenine and guanine are relatively large molecules whereas thymine and cytosine are smaller. For a symmetrical structure, large must join with small, so adenine is invariably joined with thymine and cytosine with guanine. The sequence of bases on any one chain varies as in Fig. 1.2 but their order automatically determines that of the opposite chain, which must be complementary.

The transmission of genetic information from the DNA strands is through ribonucleic acid (RNA). This is similar to DNA but it consists of one strand instead of two (Fig. 1.2), and within each nucleotide the sugar is ribose instead of deoxyribose and the base thymine is replaced by uracil. It has now been established that in the presence of the appropriate enzyme systems, RNA is formed by the splitting of a strand of DNA. It follows that only one of the two DNA strands can be copied at a time, the polymerase enzyme associated with a particular RNA sequence determining which one it shall be.

As was indicated earlier, an essential feature of any hereditary mechanism is that it must be perpetuated from generation to generation; in other words, it must be capable of replication. As a result of the distinguished work of A. Kornberg and others, we now know that the two strands of the DNA molecule are capable of peeling apart rather like the action of a zip-fastener. Moreover, provided the

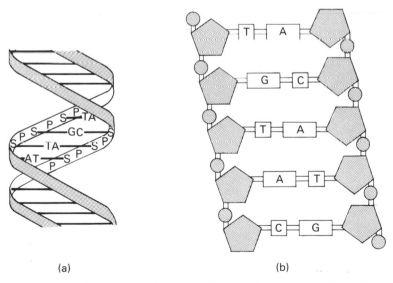

(a)                                    (b)

**Fig. 1.2** Diagrammatic representation of the linkages in a portion of a DNA molecule: (a) part of the double helix, (b) a short length of the DNA strand untwisted to show the positions of the bases. Note that the sequence in one helix is complementary to that in the other

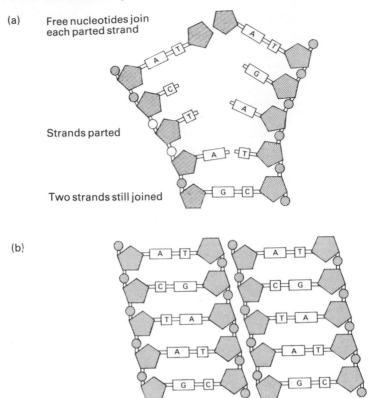

**Fig. 1.3** Replication of DNA: (a) parting of two DNA strands, (b) two new DNA strands formed. The formation of the new strands is determined by the relationship between the bases: adenine must join with thymine and cytosine with guanine

appropriate nucleotides are present in the free state, these can align themselves in an order corresponding to those of each individual strand to form a new DNA helix (Fig. 1.3).

How DNA and RNA molecules function as means of storing, transmitting, and perpetuating genetic information need not concern us here; this will be considered further in Chapter 4. At present our interest is rather in the fact that such mechanisms exist, for without them, no structure, cellular or otherwise, can be regarded as alive.

## The first protocells

Somewhere between 3000 and 4600 million years ago the ancestral cells appeared. What were these like and how were they formed? As might be expected, this has been a fertile field for speculation and attempts at laboratory simulation. In the 1950s, Oparin [1.4] and others showed that suspensions in water of colloidal particles of protein such as gelatin and various polysaccharide sugars, when treated appropriately (for instance by adjusting the pH), could be induced to clump together to form cell-size accumulations (referred to as **coacervates**). In the presence of particular enzymes, some proteins were found to undergo chemical interaction with a resulting increase in coacervate size and even a tendency to divide. Again, other coacervate systems, when provided with chlorophyll, proved capable of electron transfer from one dye to another, simulating an aspect of photosynthesis.

The coacervate model of a protocell undoubtedly proved useful in its day in showing how organic molecules of a wide variety of kinds, once formed, were capable of coming together as aggregations in many different combinations and sizes. However, Folsome [1.1] has posed one major objection to this hypothesis for the origin of life, namely that the production of coacervates in the laboratory requires a relatively high starting concentration of polymers. But, as we have already seen, in the dilute primeval 'soup', all the evidence suggests that even the simplest organic compounds must have been present in relatively low concentrations.

More recent findings suggest that the concentration gap between what was required for cell formation and what actually existed in the past is not insuperable. Evidently, many organic molecules such as those formed in the spark discharge experiments described earlier, when present together on the surface of water, tend to aggregate and form a scum. This capacity for self-assembly is widespread and does not depend on high initial concentrations. It is therefore not difficult to envisage a situation in which the accumulation of these organic microstructures reached a point where they were heavier than water and sank to the bottom of the primeval ponds. Here, they would have gained protection from the damaging ultraviolet light and the protocells could have gradually developed life-like characteristics.

## The origin of replication

It is one thing to speculate on the possible origins of the first protocells through the random aggregation of molecules but it is a jump of a different order to explain how the capacity to replicate came about. In our present state of knowledge any hypothesis must remain largely guesswork.

We have seen how the essential ingredients of nucleotides – bases, sugars, and phosphates – could well have existed in the primeval 'soup', but at a low level of concentration. This suggests that any self-replicating polymer would be likely to have existed as a single strand resembling RNA rather than DNA. The coalescing of two strands to form a helix probably represented a later stage of molecular evolution. Moreover, it seems probable that the early polymer chains must have been quite short consisting, perhaps, of ten units or less. Folsome [1.1] has suggested that these may have been arranged in a circular manner so that no free hydroxyl or phosphate groups would have been exposed to random breakdown as would have been the case in a linear arrangement. In such stabilised conditions, free bases exposed on the circle would have been able to form a series of nucleotides according to the rules of chemical pairing. In the course of time these nucleotides provided the templates for complementary copies of themselves. Whether these events could have happened without the mediation of the necessary enzymes is, again, a matter for speculation.

There is also the important question whether the origin of self-replicating polymers occurred more than once. Are the DNA and RNA systems as we know them today the only forms of polymers capable of copying one another that exist or have existed in the past? A clue to the answer to this question is provided by the asymmetry of many organic molecules such as those of sugars and amino acids. These can exist in two forms, one being the mirror image of the other (**enantiomers**) and are detected by their effect on a beam of polarised light when passed through a solution. If equal quantities of the two forms are present, the light passes through unchanged. But if there is a preponderance of one or the other the light will be rotated either to the left (laevorotatory) or to the right (dextrorotatory). We can therefore speak of L- or D-compounds depending on their symmetry as detected by the plane of polarised light. The principle is illustrated for the amino acid alanine in Fig. 1.4.

Since the origin of all living things seems to have derived from a sequence of more or less random events, we might have expected to find both D- and L-forms of such important molecules as amino acids and sugars. But this is not so. It is invariably true that all known plants, animals, and microorganisms use only D-sugars and L-amino acids. Since the alternative combination would appear to have been just as effective, why has only this particular arrangement survived? There are two possible answers. First, the universality of the situation suggests that life only evolved once. Second, the particular enantiomers used were in some way preferable to the others for performing the range of chemical reactions that represent the life processes. Support for this view has been forthcoming from recent work on the rates of polymerisation of amino acids, where it has been found that L-compounds polymerise the more readily of the two. We are still a long way from understanding the precise significance of these findings in biological terms, but the fact that particular molecules occur universally in living systems suggest that they may be better suited to certain requirements and that their survival was not just a matter of luck.

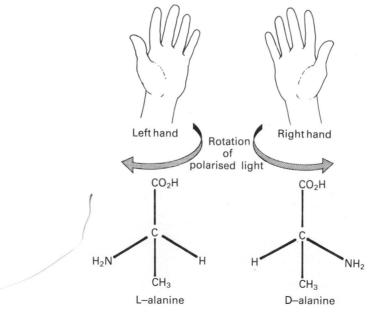

**Fig. 1.4** Enantiomers as mirror images, illustrated by the amino acids L- and D-alanine

# The origin of prokaryotes

The difference between the essentially non-living protocells and living cellular structures represents another enormous evolutionary leap. Moreover, it brings us into the sphere of organisms that still occur today. In order to appreciate the significance of the different levels of cell organisation that now exist, it is important first to outline the main groups of living organisms.

The largest classificatory groupings of living things are kingdoms of which there are usually considered to be five – Animalia (higher animals), Plantae (higher plants), Fungi, Protista (unicellular organisms such as *Amoeba*, which may also exist as colonies), and Monera (minute single-celled organisms including bacteria and blue-green algae). However, this classification serves to conceal one of the most fundamental differences, namely the distinction between monerans and the rest. This is based on the type of cells that they possess. Thus, although cells of the four higher kingdoms are grouped together as **eukaryotes**, those of the Monera are **prokaryotes**. The characteristics of the two groups are compared and contrasted in Table 1.2 and Fig. 1.5. It will be seen that in every way prokaryotes are simpler in their organisation and less specialised than eukaryotes. In studying the possible origins of life, we therefore need to concentrate our attention in the first place on the distribution of prokaryotes in space and time, and on possible ways in which these organisms might have been derived from their protocell ancestors.

During the 4600 million years of the Earth's existence, the remains of multicellular plants and animals (eukaryotes) are not detectable as fossils until comparatively recently – about 700 million years ago (see Chapter 2). Unicellular

**Table 1.2**  Comparison of the characteristics of prokaryotic and eukaryotic cells

|  | *Prokaryotic cells* | *Eukaryotic cells* |
| --- | --- | --- |
| Cell size (μm) | 0.5–2 | 5–20 |
| Cell nucleus | Nucleoid with no boundary membrane; no chromosomes; chromatin in form of chromoneme | Nucleus bounded by nuclear membrane; contains chromosomes |
| DNA | No protein covering; ends of helix joined to form loop | Double helix surrounded by protein sheath; ends of helix free |
| Cell division | No mitosis or meiosis; division by simple fission | Mitosis and meiosis |
| Reproduction | Chromonemal material passed from one cell to the next in the process of fission | Usually sexual by haploid gametes forming diploid zygotes |
| Motile structures | No true cilia or flagella | True cilia and flagella may be present (e.g. in Protista) |
| Organelles in cytoplasm | No mitochondria or chloroplasts; oxidase enzymes, when present, free in cell; many anaerobic; pigments, when present, free in cell | Plants have chloroplasts; all have mitochondria carrying oxidase enzymes |

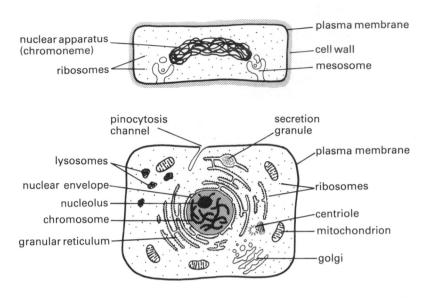

**Fig. 1.5** Schematic diagrams of (a) a prokaryotic cell, (b) a eukaryotic cell

protists may well have existed another 300 million years earlier. But the remains of prokaryotes have been identified as far back as 3000 million years. We are thus able to pinpoint the origin of life on the Earth as having occurred some time during the first 1600 million years. Moreover, as we have already seen (p. 7), there is considerable evidence to suggest that living processes evolved only once. As to the first minimal prokaryotic cells, we can only hazard a guess as to their structure. They must have included one or more polymers consisting of a short chain of nucleotide-like units capable of replication. As was suggested earlier, pairs of chains may well have been joined together at their ends to form rings thereby achieving greater chemical stability. It is significant that this is precisely the situation that we find in prokaryotes today. A prime requirement for life is the availability of food and the capability to utilise it as a source of energy and to facilitate growth. The absence of oxygen in the atmosphere would have necessitated a form of anaerobic respiration resembling fermentation, and this would have made use of some of the wealth of organic material that must have been available.

In the course of time, some forms may well have evolved the capacity to fix atmospheric carbon dioxide and to form organic compounds such as sugars, just as many living species do today. Once the appropriate pigments had been produced by incorporating metal ions such as magnesium, photosynthesis would have been possible. At first, the process was no doubt anaerobic, as occurs in some living bacteria. With the development of aerobic photosynthesis with oxygen as a by-product, anaerobes will have been at a disadvantage. Their distribution must gradually have become restricted to oxygen-free environments such as mud in which iron, sulphur, and other elements formed compounds, which could have been used as raw materials for respiration. Some bacteria survive today in such environments.

## The origin of eukaryotes

Reference to Table 1.2 shows that some of the differences between prokaryotes and eukaryotes are fundamental. In particular, the question arises as to how organelles such as mitochondria and chloroplasts came to be incorporated in the cytoplasm of all eukaryotic cells. Another equally important question is how the complex process of sexual reproduction first arose.

Mitochondria evidently occur in the cells of all living organisms except bacteria and blue-green algae. They are extremely variable in size and appearance, some being sausage-shaped, others more slender and rod-like. Their length ranges from about 1.5 µm to 10 µm. Their internal structure is remarkably constant. Each mitochondrion is bounded by two membranes, the outer being smooth and independent of other structures. The inner membrane differs structurally and chemically. It is folded in an elaborate way forming a series of shelf-like pockets known as **cristae** which present a relatively large surface area (Fig. 1.6). Mitochondria can synthesise various enzymes that are particularly concerned with oxidation and energy exchange, the surface of the cristae providing a ready means of promoting metabolic activity. In a sense they can be regarded as independent living entities. This view is reinforced by the fact that mitochondria contain their own DNA which closely resembles that found in prokaryotic cells such as bacteria, in that the ends of the strands are joined to form loops. Mitochondria also

**Fig. 1.6** Electron micrograph of mitochondria in insect flight muscle. Note the folding of the inner membrane to form cristae

possess ribosomes that are similar to those in prokaryotes but are considerably smaller than those in eukaryotes.

Valuable clues to the ancestry of mitochondria have also been obtained from laboratory experiments on the sensitivity to various antibiotics of the protein-synthesising systems in mitochondria, bacteria (prokaryotes), and the cytoplasm of eukaryotes [1.5, 1.6]. These are summarised in Table 1.3. Clearly there is evidence of a similarity in the protein-synthesising systems of mitochondria and bacteria (prokaryotes), both of which differ from those of eukaryotic cytoplasm.

**Table 1.3**   Sensitivity to antibiotics of the protein-synthesising systems of mitochondria, prokaryotes, and eukaryotes (after Tribe, Morgan and Whittaker 1981)

| Antibiotics | Mitochondria | Bacteria (prokaryotes) | Cytoplasm (eukaryotes) |
|---|---|---|---|
| Anisomycin | − | − | + |
| Cyclohexamide | − | − | + |
| Emetine | +/− | − | + |
| Mikamycin | + | + | − |
| Chloramphenicol | + | + | − |
| Erythromycin | +/− | + | − |
| Lincomycin | +/− | + | − |

+, Inhibition; −, no inhibition; +/−, mitochondria from some species inhibited but those from others are not

All these data strongly suggest that in their early days mitochondria may have been free-living prokaryotes. They may well have become incorporated in the cytoplasm of a primitive eukaryote by some process resembling **endocytosis**, in which a particle is ingested by a portion of the cell membrane flowing round it like the feeding of an *Amoeba*. The outer membrane of the mitochondrion could thus represent that of the original prokaryote that engulfed it. It has been suggested that such an arrangement survived because the two organisms were able to exist closely together (**symbiosis**) – a form of relationship that is widespread among plants and animals today. Thus, the familiar lichens (Fig. 1.7) consist of a symbiotic association between a fungus and unicellular algae, and many animals house bacteria or protozoa in their guts, which aid digestion, while obtaining their food from the environment provided by their hosts.

But what were the benefits of symbiosis gained by the early eukaryotes and their partners? All eukaryotes are aerobic but many prokaryotes are anaerobic. Indeed, in primeval conditions anaeroby will have been the predominant mode of life. One of the major problems facing the early eukaryotes must have been the need to adjust to an effective utilisation of oxygen as the proportion of the gas in the atmosphere increased. Any bacterium-like organism possessing a form of metabolism even remotely resembling that of a mitochondrion would have been an asset as a symbiont in such conditions. No doubt the early enzyme systems were few and limited in their action so that such symbiosis as occurred was sporadic and reversible, and potential symbionts could also have been capable of a free-living

**Fig. 1.7** A typical lichen (a symbiotic association between a fungus and algae), *Evernia prunastri*, growing on an oak

existence (**facultative symbiosis**). Later, in the face of increasing need, the association could well have proved indispensable to the survival of both partners and so became irreversible (**obligatory symbiosis**). The symbiont prokaryotes thus became permanently housed within the cytoplasm of the eukaryotic cell from which they would have derived protection from the external environment and a supply of food.

The sequence of events leading to green plants may have been somewhat similar to that of the evolution of mitochondria. In those prokaryotes that contain photosynthetic pigments, such as the blue-green algae, the pigments invariably occur free in the cytoplasm. But in all green plants, the photosynthesising pigment chlorophyll is contained within **chloroplasts**. Chloroplasts closely resemble mitochondria in their general composition: thus, each is bounded by a double membrane; the outer one is smooth and independent of other structures whereas the inner membrane is much folded and presents an appearance even more complex than that of a mitochondrion (Fig. 1.8). The inner portion of the chloroplast is divisible into a colourless **stroma** and a series of disc-like structures (**thylakoids**) that contain chlorophyll and are arranged one upon another in small stacks (**grana**). Chloroplasts are usually larger than mitochondria (their length ranges from 4 to 10 μm). Like mitochondria, they contain DNA which is capable of replication. Not only are they the sites of photosynthesis but they are also able to carry out phosphorylation whereby sunlight is used to synthesise ATP, a process resembling oxidative phosphorylation in mitochondria.

Chloroplasts could thus have been derived by symbiosis from free-living prokaryotes in a manner similar to mitochondria. The symbiotic theory is supported by the composition of the RNA and proteins of chloroplasts and

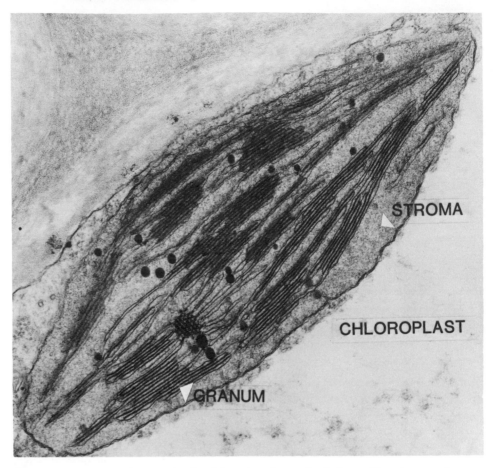

**Fig. 1.8** Electron micrograph of a chloroplast from a leaf of pea. Note its complex internal structure and the grana containing chlorophyll

mitochondria. In mitochondria the RNA and proteins resemble those of bacteria, not the surrounding cytoplasm. Similarly, in chloroplasts they are different from the rest of the cell but are closely similar to those in the prokaryotic blue-green algae.

## The origin of sex

Among the early prokaryotes, as still happens today, multiplication no doubt took place by the simple process of splitting in two (**fission**), the two daughters being identical both to one another and with their mother cell. Sexual reproduction must have evolved among eukaryotes where it is now almost universal. In a sense, the two processes are opposites because asexual reproduction involves the division of single cells, whereas sexual reproduction consists of the fusion of two cells (**gametes**) to form one cell – the fertilised egg (**zygote**). The two situations are contrasted in Fig. 1.9. Among organisms living today, such as the simplest

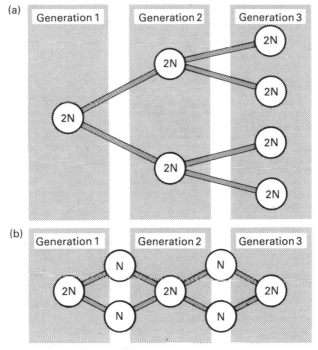

**Fig. 1.9** Patterns of (a) asexual and (b) sexual reproduction in a diploid organism (2*N* chromosomes)

aquatic algae, it is quite easy to construct a hypothetical series to show the steps by which sexual reproduction might have come about. Even prokaryotes such as the bacterium *Escherichia coli* possess a primitive form of sexuality: when *E. coli* cells collide, as in a culture medium, a bridge may form between two adhering **conjugants** and material, including DNA carrying genetic information, then passes from one cell to the other. This is sexual reproduction at its simplest, the two cells each performing the dual functions of non-sexual (**somatic**) and sexual activities. In all higher eukaryotes the development of sexual reproduction has been taken two steps further:

(i) In the differentiation of the sexual and somatic parts of the body.

(ii) In the separation of the sexual side into male and female. This usually involves marked differences in the two kinds of gametes (**anisogametes**), the male sperms being numerous, small and mobile, and the female eggs, which usually contain stored food, being fewer, larger and generally immobile. In some of the simplest plants the two kinds of gametes are indistinguishable in appearance (**isogametes**) but they nonetheless show a distinct polarity in their tendency to fuse together (**syngamy**), as do sperms and eggs. By definition, all gametes are **haploid**; that is to say they contain half the number of **chromosomes** (*N*) that are found in **diploid** (2*N*) somatic cells. The halving of the chromosome number usually occurs during gamete formation (**gametogenesis**) and results from a kind of cell division known as **meiosis** which is not found in prokaryotes (see Table 1.2, p. 9).

At first sight, it seems extraordinary that the complex process of sexual reproduction ever came into being. Judged in purely quantitative terms, it is less efficient than asexual reproduction because two cells are needed to create one. Presumably it must have conferred on the individual certain other advantages that were unattainable through asexual multiplication. But what could these advantages have been? In attempting to answer this question it is important to distinguish between the long-term implications of the two reproductive processes and possible benefits for the individual organism at the time.

Concerning the implications, we must bear in mind that all organisms live in a changing world and as their environment alters, they need to be flexible enough to adjust accordingly if they are to survive. In order to achieve this flexibility living things must vary. **Mutations** – inherited changes occurring at particular gene sites – provide the source of this variation. Now, in asexual reproduction and in the absence of mutation, all the descendants of one individual are genetically identical (**clone**). When a mutation occurs this will be duly passed on through successive divisions, but its dispersal within a population is bound to be slow with an inevitable limitation on overall variation. In sexual reproduction, each offspring carries a unique combination of genes resulting partly from the reshuffling process that takes place at meiosis during gametogenesis and partly from the random union of pairs of gametes. It follows that sexually reproducing populations must be more variable than asexual populations and therefore able to achieve greater flexibility in adjusting to a changing and diverse environment. However, although the long-term advantages of these characteristics for **populations** may be self-evident, it is less easy to see how they could have helped the **individual** in the short term except on the rare occasions when a particular mutation happened to fit a particular environmental requirement. So we must look elsewhere for a more plausible explanation of the development of sex.

One factor that may have played an important part in the origin and elaboration of sex derives from the female gamete. Two of its essential features are a store of food available to the developing embryo and a tough outer case that provides protection against adverse conditions. Many plants and animals alive today overwinter or combat such unfavourable circumstances as cold and drought in the form of zygotes, fertilised eggs or as seeds with protective outer coats. Such species as aphids (greenfly) reproduce both asexually and sexually. Whether such situations are remnants of a primitive condition or secondary specialisations need not concern us here. But it is significant that when circumstances are favourable and food and other resources abundant, asexual multiplication is usually the rule; as conditions deteriorate – for instance with the approach of winter – sexual reproduction generally takes over. It may well be, therefore, that the origin of sex was closely correlated with the need to protect an increasingly elaborate and sensitive embryonic organism during periods of adversity.

## Summary

1  Modern studies have established that the age of the Earth is far greater than was supposed in the eighteenth and nineteenth centuries. It is now estimated at about 4600 million years.

2  Radioactive isotopes have played an important part in dating different events in the Earth's history. Carbon-14 breaks down relatively quickly and its limit

as a geological clock is about ten half-lives, or 57 300 years. For estimating longer periods, uranium-238 and 235, also thorium-232 are frequently used.

3   From the composition of ancient rocks we can deduce a good deal about the Earth's primeval atmosphere. Evidently, there was no oxygen and the average temperature was about 25 °C.

4   Available evidence suggests that in its early years there was much volcanic activity on the Earth's surface and thunderstorms were frequent.

5   Attempts to simulate these conditions in the laboratory have shown that a wide range of organic compounds could have been formed from the gases present. Some of these products, such as amino acids, would have had potential biological importance.

6   Two essential requirements of the first cells were the concentration and assembly of organic molecules in an orderly manner, and the capacity for self-replication.

7   Spark discharge experiments have shown that the essential ingredients for the formation of primitive nucleotides could have been formed in the conditions that prevailed on the primeval Earth.

8   Various models have been put forward to account for the development of the first protocells and the origin of replication. The fact that all organisms living today contain D-sugars and L-amino acids suggests that life evolved only once.

9   There are fundamental differences between the characteristics of prokaryotic and eukaryotic cells. Since the remains of prokaryotes have been identified in deposits 3000 million years old, life must have evolved during the previous 1600 million years. The first eukaryotic fossils are dated at 700 million years although unicellular forms may have occurred about 300 million years earlier.

10   In tracing the origin of eukaryotic cells, there is evidence to suggest that mitochondria and chloroplasts developed from prokaryotes such as bacteria or blue-green algae that became symbiotic.

11   In the development of eukaryotes, an important landmark was the evolution of sex.

12   Although sexual reproduction leads to increased variation within a population, this long-term benefit could have played little or no part in promoting its origin. A factor which could well have conferred increased survival on its possessors is the protection against adverse conditions afforded to the developing embryo by the thick wall and food reserve of the fertilised egg or zygote.

## Topics for discussion

1   What are the principal limitations of the various attempts to simulate the early conditions on the Earth and to synthesise the compounds essential to life?

2   What criteria could be used to decide whether protocells are living or not?

3   It has been suggested that in their early days, mitochondria and chloroplasts may have been free-living prokaryotes. What evidence is there for this idea?

4   One theory of the origin of life on Earth claims that it originated not on the Earth's surface, but from extraterrestrial space. On this hypothesis, what kinds of bodies might we expect to have been the first arrivals?

5   What evidence is there that symbiosis may have played an important part in the evolution of eukaryotic cells?

6   What factors could have prompted the evolution of sex? What was its significance? Is it possible to construct a hypothetical sequence of events leading to the development of heterogamy and fertilisation?

7   To what extent has the origin of life been linked to the evolution of the Earth's atmosphere?

8   Why is the continued origin of life on Earth not occurring now?

## References

1.1   Folsome, C. E. (1979) *The Origin of Life*. W. H. Freeman, San Francisco

1.2   Miller, S. L. (1953) 'Production of aminoacids under possible earth conditions'. *Science* **117**, 528

1.3   Watson, J. D. (1968) *The Double Helix*. Weidenfeld and Nicolson

1.4   Oparin, A. I. (1956) *The Origin of Life on Earth*. Oliver and Boyd

1.5   Tribe, M. and Whittaker, P. (1972) *Chloroplasts and Mitochondria* (Studies in Biology, No. 31). Edward Arnold

1.6   Tribe, M. A., Morgan, A. J. and Whittaker, P. A. (1981) *The Evolution of Eukaryotic Cells* (Studies in Biology, No. 131). Edward Arnold

# 2

## Geological evidence for change

One of the greatest problems in studying the evolution of living things is the lack of tangible evidence for change. As we saw in the last chapter, the remains of prokaryotes have been found in deposits as old as 3000 million years but these are exceedingly fragmentary. The first identifiable eukaryotic fossils (aquatic algae) date to around 700 million years. In other words, widespread tangible evidence for the development of life is only available for about 15 per cent of the Earth's existence. But relatively recent geological history also has its limitations, as Darwin fully realised. Writing in *The Origin of Species* (Chapter 9) on the imperfections of the geological record, he observes,

> I look at the natural geological record as a history of the world imperfectly kept, and written in a changing dialect; of this history we possess the last volume alone, relating only to two or three countries. Of this volume, only here and there a short chapter has been preserved; and of each page, only here and there a few lines. Each word of the slowly-changing language, in which the history is written, being more or less different in the successive chapters . . .

Since Darwin's time our knowledge of the worldwide fossil record has greatly improved, while methods of study and description have become more standardised. But even so, the panorama is still discontinuous with relative completeness in some places and tantalising gaps in others. In order to judge these in perspective we must first know something of the nature of fossils and the circumstances in which they are formed.

### What are fossils?

Fossils are the dead remains of plants and animals preserved in rocks. They are usually the hard portions such as the chitinous exoskeleton of arthropods, the calcareous coverings of echinoderms, or the bony endoskeletons of vertebrates. Plant fossils include woody tissues, leaves, seeds, even pollen. Sometimes, fossils occur as moulds of a hard structure such as an animal limb on the surrounding rock. Such imprints can be of many different kinds such as footsteps, burrows, or the borings of bivalve molluscs such as the piddock, *Pholas*, into soft sediments.

As was indicated earlier, the occurrence of fossils in different strata of the Earth's surface tends to be discontinuous. To some extent this represents the abundance of organisms when a particular sediment was laid down. But it also depends upon the composition of the surrounding medium [2.1]. For dead material to be fossilised successfully the following requirements must be satisfied:

(i) Coverage must be rapid in order to avoid destruction by predators or by the physical environment such as wind and rain.

(ii) Decomposition by microorganisms such as bacteria must be slowed down or stopped altogether.

(iii) The surrounding medium must be of a kind to permit the originally living material to be impregnated by chemicals in solution such as silica or calcite, a process known as **petrification**. As a result, the weight and hardness of the fossil become greatly increased. Alternatively, chemical changes may occur in the material itself leading to **recrystallisation**. This seldom affects the appearance of the structures concerned but, in time, their original components may be completely replaced. The consequence of this process of **replacement** can be very beautiful, as in the preserved remains of ferns and horsetails found in the Coal Measures (Carboniferous) where the tissues have been converted into a film of carbon (**carbonisation**) as is illustrated in Fig. 2.1.

Most fossils occur in sedimentary rocks and, given the peculiar requirements for preservation, it is not surprising to find some media are suitable whereas others are not. Thus limestone, chalk, and clay frequently contain many fossils whereas sands and gravels, being more porous, are less productive. One of the reasons for the poor fossil record of man is that human burials have usually been in ground which can be easily worked and is therefore well aerated. Decomposition by bacteria and other organisms has, therefore, been rapid. In relatively recent times, a striking example of human preservation has been provided by Bronze Age burials in Denmark around 3000 BC. The Danes buried their dead in coffins hewn out of tree trunks and lowered them into peat bogs where the wet, acidic

**Fig. 2.1**　Fossil fern (*Neuropteris*) from the Carboniferous. The detailed structure is beautifully preserved

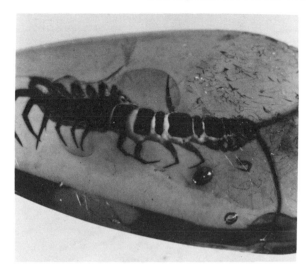

**Fig. 2.2** Myriapod preserved in amber for about 100 million years

conditions of the anaerobic mud inhibited bacterial action. Not only are the bodies well preserved but so too are such materials as hair, clothing, and leather footwear.

Some of the most perfect examples of preservation occur in the gum of trees which is sticky and exudes from damaged surfaces. Leaves, pollen, and small insects such as ants tend to adhere to it like a flypaper and become enveloped. Amber is the fossilised remains of tree resin and the organisms it contains often exhibit almost as much detail as when they died millions of years ago (see Fig. 2.2).

## The geological record of change

The greater part of the Earth's crust is composed of rocks which have been laid down one layer upon another as a result of the physical forces of erosion of which water, wind and temperature are the most important. Such **sedimentary** deposits thus represent a sequence in time and the fossils they contain present a changing picture of life in so far as preservation has occurred.

### Cuvier and the theory of catastrophes

In fact, the geological situation is a good deal more complex than appears at first sight, and we are far from understanding the full significance of the data from geology and palaeontology that are still being accumulated. One of the first men to carry out systematic research in this field was the Frenchman George Cuvier who published his findings from the rocks in the vicinity of Paris in 1812. He was struck by the great differences, both qualitative and quantitative, between the fossil content of the different strata. Thus, at one level there were the remains of marine animals, at another freshwater species and in yet another, no fossils at all. In modern terminology, he had demonstrated the phenomenon of **discontinuity** which we now accept as a feature of geological deposits worldwide. The problem facing Cuvier was how to interpret this discontinuity. In accordance with the religious beliefs of his time, he believed in the literal truth of the Book of Genesis. For him, the absence of a group of organisms from the fossil record meant their **extinction**. This he explained in biblical terms as being due to a series of catastrophes at different stages in the Earth's history, the latest being the flood recorded in Genesis. After each catastrophe, which occurred at varying times in different places, recolonisation by plants and animals took place, partly through further creation and partly through the immigration of species from elsewhere which had escaped destruction.

Cuvier's outlook was essentially anti-evolutionary and grounded in a belief in the fixity of species. On the other hand, he accepted the evidence from his own data on discontinuity that after each catastrophe the organisms that followed tended to be more advanced than those before them. This idea of **progressionism** could be regarded, in a sense, as a first step towards the more fluid concept of evolution. Moreover, in accepting progressionism, Cuvier also laid the foundation of the concept of **radiation**, for he realised that over great periods of time organisms had diverged into a number of different groups some of which are still accepted as part of the classificatory system today. Such a view differed fundamentally from current doctrine, which was still based on the *Scala Naturae* (ladder of life) of Aristotle (p. 117) and envisaged all living things as occupying a single graded series from the simplest at the bottom to the most complex (i.e. man) at the top.

### Lyell and the principle of uniformity

Among the luggage that Darwin took with him on the voyage of the *Beagle* (p. 23) was the first part of a three-volume work, *Principles of Geology*, by Charles Lyell (1797–1875) published between 1830 and 1833 (the two other volumes reached him later when he was in South America). It was presented to him by his friend Professor J. S. Henslow who is said to have instructed him to read it 'but by no means to believe'. Darwin not only read the book but believed what Lyell had said. Indeed, the principles it enunciated and the evidence it contained were to play a major part in convincing him that evolution had occurred in the past. This was later reinforced by evidence collected in South America and elsewhere by Darwin himself.

Lyell's work brought a new precision into the study of geology. His careful analysis of fossil succession (or the lack of it) in different parts of the world, also the forces concerned in producing stratification of the sedimentary rocks revealed the structure of the Earth's crust in a new light. In particular, Lyell pointed out that the forces responsible for laying down successive strata were still acting today in a precisely similar manner to the past. There was therefore no need to postulate catastrophes. On the contrary, geological succession could be explained as a universal process, working continually over time and being brought about by the combined action of **erosion** and **sedimentation**. This revolutionary approach came to be known as the principle of **uniformity**. It at once made sense of the fossil record, enabling valid comparisons to be made of the organisms found at different levels in comparable strata. Lyell had no precise means of determining the age of the Earth, but from a rough estimate of the rate of deposition of different sediments he concluded that it was far greater than had previously been supposed.

Moreover, the constitution of the Earth's surface was not stable but in a constant state of change.

Looking back on the contributions of Lyell, it is something of a paradox that he used his theory of uniformity to argue not only against catastrophism but also against the idea of evolution. He viewed geological systems and the physical environment as in a constant state of change; in this he was correct. But he was opposed to progressionism and regarded the biotic (living) environment as unchanging. Like Cuvier, he considered species to be immutable and disappearance from the fossil record to indicate extinction, any gaps being filled by forms that had survived or by the creation of new but similar ones. As far as Darwin was concerned, Lyell's views were of great importance in establishing the facts of change but of little significance in suggesting how that change could have come about.

## Darwin and the geological record

Charles Darwin (1809–82) sailed from Plymouth in HMS *Beagle* on 27 December 1831, and returned to Falmouth after his epic 5-year journey round the world on 2 October 1836. The circumstances in which he came to join a naval ship's company as unpaid naturalist and the precise route taken by the *Beagle* in the course of its surveying operations need not concern us here. These have been fully documented and analysed elsewhere [2.2]. Suffice it to add that whereas Darwin departed from England a confirmed creationist in outlook, he was to return 5 years later firmly convinced that evolution had occurred, although he was still uncertain about the mechanism involved.

During his voyage round the world, Darwin had unique opportunities of making geological observations of all kinds, ranging from the formation of coral reefs to variations in rock stratification and the effects of movements in the Earth's surface. Some of his most significant findings were in different areas of South America. Thus, on 20 February 1835, he recorded his sensations during an earthquake at Concepcion Bay in southern Chile. On making measurements a few days later he found that the land had risen about a metre. Nearby, putrefying mussels were found still clinging to rocks now some 3 m above high-water mark. Again, not far away he found fossilised conifer trees at a height about 2500 m (7000 ft) above sea level. The fact that they were projecting almost perpendicular from the stratum in which they were embedded indicated that they must have grown there and had not been washed in from elsewhere. The beds containing the fossils were covered by thousands of metres of alternating volcanic ash and sediment, showing that at one time the trees had been buried deep in the Atlantic Ocean which extended westwards. Since then they must have been raised some 2500 m above sea level.

From his observations in Concepcion Bay, Darwin concluded that the elevation of land could take place far more quickly than had previously been imagined. Near Callao in Peru, he found the remains of human activity in the form of plaited rushes and cotton string, now at a height of about 25 m (85 ft) above sea level. Such evidence of colonisation by man indicated that elevation must have taken place since the time when the Indians inhabited Peru about 150 years earlier. Again, near Valparaiso (Chile) Darwin showed that an estimated rise in land level of about 3 m had occurred over a period of 17 years. Such observations raised the

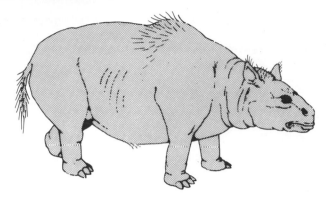

**Fig. 2.3**   Reconstruction of *Toxodon*, a huge rhinoceros-like animal which Darwin found as a fossil in South America

precision of geological observation from a purely descriptive approach to a more objective and quantitative one.

In addition to his studies of elevation, Darwin accumulated a large amount of evidence regarding fossil animals occurring in different strata. Included among these were the remains of many mammals such as monkeys, rodents, and ruminants. Some were of vast size such as the rhinoceros-like *Toxodon* (Fig. 2.3) which possessed large, curved upper teeth for gnawing. Many of the fossils bore a striking resemblance to species still alive although they differed from them in detail, particularly size. From this, Darwin concluded that similar species, although distinct, must be closely related. Evolutionary **divergence** must have occurred comparatively recently in geological time.

Yet, in spite of this spate of discoveries, Darwin remained well aware of the overall imperfection of the fossil record as it was then known. In his conclusions to *The Origin of Species* (Chapter 15) he observes,

> The noble science of Geology loses glory from the extreme imperfection of the record. The crust of the earth with its embedded remains must not be looked at as a well-filled museum, but as a poor collection made at hazard and at rare intervals. The accumulation of each great fossiliferous formation will be recognised as having depended on an unusual occurrence of favourable circumstances, and the blank intervals between the successive stages as having been of vast duration.

### The geological record since Darwin

One of the problems facing Lyell in his attempts to introduce a more objective approach into the study of geology, was the lack of any precise method of dating rocks and therefore of judging the rates at which they were laid down. His doctrine of uniformity postulated that the forces responsible for shaping the Earth's crust were the same the world over but that they operated in different places at differing rates. Lyell could do no more than estimate these rates in general terms. Darwin took his approach a step further while in South America and in a few instances was able to measure the rate of land elevation using known geological or historical events as markers. As we saw in Chapter 1 (p. 1), the use of radioactive isotopes has opened up a new dimension in geological dating, and we now know that the origin of the Earth and the development of life upon it are far more ancient than even Lyell believed.

## The fossil sequence

As we saw in Chapter 1, modern estimates suggest that the age of the Earth is about 4600 million years. Of the first 1600 million years of its existence we know little and fossil records are totally absent. The first remains of prokaryotes have been identified in deposits about 3000 million years old, while the earliest eukaryotic fossils are dated at around 700 million years. This vast span of time extending backwards from some 600 million years ago is referred to as the Precambrian (Table 2.1).

**Table 2.1**   Summary of events during the Precambrian

| Time from the Earth's origin (millions of years) | Evidence of life |
| --- | --- |
| 1600 | No fossil records |
| 3000 | Remains of prokaryotes |
| 3900 | First eukaryotes |

By the time the Precambrian ended, prokaryotes in the form of bacteria and blue-green algae (also called cyanobacteria) were well established, as were simple multicellular algae (eukaryotes) resembling primitive seaweeds. Coelenterates with hard parts such as corals and annelid worms capable of forming resistant tubes were just beginning to appear.

The time range of the fossil record as we know it today and the still greater period that preceded it are difficult to comprehend, and this poses problems in placing the origins of existing plants and animals in perspective. E. D. Wilson and colleagues have devised a realistic method of simplification by reducing the whole of geological time to the span of a 30-day month so that one day corresponds to 150 million years [2.3]. If we assume that the first living things appeared about 4000 million years ago, this corresponds to day 4. The first prokaryotic fossils occur at day 10 and the earliest eukaryotes at day 24. Invertebrates were flourishing by day 27 and the first vertebrates appeared on day 28. Mammals occurred on day 29 and flowering plants early on day 30. On this scale, the first true men appeared about 10 minutes ago while the whole of recorded history as we know it occupies only the last 30 seconds. The scheme is summarised in Table 2.2.

A modern version of the geological time-scale is shown in Fig. 2.4; the vertical bars represent the known existence of the principal groups of plants and animals. It will be noted that only the bacteria and blue-green algae (prokaryotes), and the early algae (eukaryotes) extend appreciably beyond 600 million years. The scale for the time since the Precambrian is divided into three major **eras** – the Palaeozoic, Mesozoic, and Cenozoic, which includes the present. Eras are subdivided into **periods** of varying duration, each characterised by a particular range of geological deposits. Thus, in Britain, the Devonian is associated mainly with

**Table 2.2** Evolutionary history represented as a 30-day month (from Wilson *et al.* 1973)

| Days | Conditions of life on Earth |
| --- | --- |
| 1–3 | No life |
| 4 | First evidence of life |
| 5–9 | |
| 10 | Prokaryotes |
| 11–23 | |
| 24 | Eukaryotes |
| 25–26 | |
| 27 | Invertebrates |
| 28 | Vertebrates |
| 29 | Mammals |
| 30 | Flowering plants |
| 10 minutes | Man |
| 30 seconds | Recorded history |

sandstones, the Carboniferous with coal measures, and the Cretaceous with chalk. Constructed in this way, the time-scale provides a useful comparison of the origins and duration of the different groups of organisms, but gives no indication of their relative abundance. Once life had reached the multicellular stage in the Cambrian, rapid diversification occurred, but the curve of increasing variety and numbers has by no means always been smooth. Thus, extinction on a large scale accompanied by rapid evolutionary changes, took place towards the end of the Cambrian (extinction of many trilobites); Ordovician (first fishes); Devonian (first amphibians); Permian (extinction of many marine animals); Triassic (first dinosaurs and early coniferous plants); and Cretaceous (climax of dinosaurs and first flowering plants). The most serious upheaval occurred towards the close of the Permian when no fewer than twenty-four orders and about half the families of animals throughout the world became extinct. The causes of these catastrophic events and their occurrence on such a vast scale still remain obscure. Some geologists postulate major land changes due to elevation with consequent drastic alteration in water levels; others believe that changes in climate occurred and the Earth became much cooler. Others again, claim that the atmosphere was subjected to massive doses of cosmic radiation from outer space. The truth is that what was probably the greatest crisis ever to strike living organisms still remains unexplained.

## The Quaternary period in Britain

Reference to Fig. 2.4 shows that the most recent geological period, the **Quaternary**, is so short (less than 2 million years) that it hardly features on a time-scale spanning 700 million years. Yet, the events that took place then are of particular significance, for it was as a result of their influence that the ecology of Britain, as we know it, exists today. Beyond historic times, the precise dating of events is uncertain and no two authorities agree on their exact nature and duration. However, there is now sufficient consensus for a generalised summary of

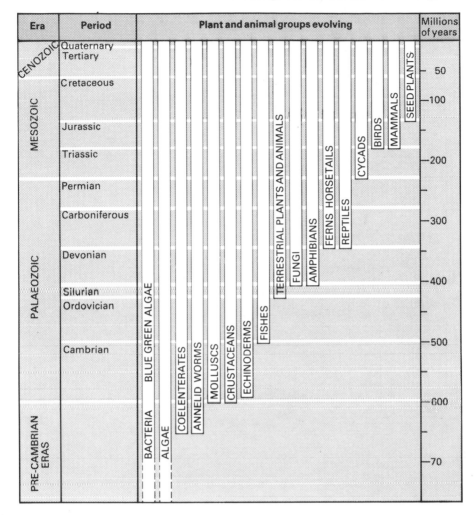

**Fig. 2.4** The geological time-scale showing the duration of some of the main groups of plants and animals

happenings during the Quaternary in Britain, and these are summarised in Table 2.3. The pattern emerging is of four ice ages (**glaciations**) of varying duration and severity alternating with warmer **interglacials**, the last of which, the **postglacial** (Holocene) period, extends to modern times. During the past 10 000 years or so, the topography of our landscape together with its characteristic flora and fauna have gradually assumed their modern form. Apart from recent immigrants, the present-day plants and animals of Britain must have been derived from:

(i) Relict species which survived the last ice age (Fourth Glacial).

(ii) Postglacial colonists from the Continent. These will have been part of the great northward movement, that took place for a period of around 4000 years between the final recession of the ice and the severing of the land bridge with England about 8000–7000 years ago.

**Table 2.3** Summary of events in Britain during the Quaternary period

| Period | Glaciation | Climatic conditions | Human remains and cultures | Characteristic fauna and flora |
|---|---|---|---|---|
| PLEISTOCENE | Postglacial | Sub-Atlantic age; climate oceanic, similar to today; about 1000 BC | Historic times; includes Iron Age | Similar to today |
| PLEISTOCENE | Postglacial | Atlantic age; wetter and warmer than today; connection with Continent severed; about 5000 BC | Includes Bronze Age | |
| PLEISTOCENE | Postglacial | Boreal age; continental climate; dry with cold winters and warm summers; about 6800 BC | | Appearance of oak and other deciduous trees |
| PLEISTOCENE | Postglacial | Pre-Boreal age; sub-arctic conditions; dry and cold; about 8300 BC | Neolithic (New Stone Age); polished implements | Pine, birch, and willow predominant trees |
| HOLOCENE | Fourth Glacial | Less intense than Third Glaciation; cold in north; tundra conditions over most of Britain | Magdalenian culture | Reindeer, arctic fox, marmot, mammoth, and most other arctic animals extinct |
| HOLOCENE | Third Interglacial | Shorter and cooler than Second Interglacial | Modern man (*Homo sapiens*); Aurignacian culture | |
| HOLOCENE | Third Glacial | Cold conditions as far south as River Thames | Mousterian culture | Mammoth common |
| HOLOCENE | Second Interglacial | Longest interglacial; relatively warm in south but cold in north | Neanderthal man; Acheulian culture | Rich fauna and flora in south |
| HOLOCENE | Second Glacial | Great glaciation, particularly in the east of England | Chellean culture | |
| HOLOCENE | First Interglacial | Relatively warm period lasting many thousands of years | | Straight-tusked elephant still in evidence |
| HOLOCENE | First Glacial | Ice sheet from Scandinavia to coast of Durham and Yorkshire | | |
| PLIOCENE | Preglacial | Sub-Atlantic conditions; similar to the present but sea much colder | | Straight-tusked elephant, hippopotamus, and rhinoceros; molluscs predominant in the sea |

# Problems of fossil continuity

As we saw earlier, Darwin took a dim view of the fossil record and its incompleteness in providing evidence for the modification of species throughout time. He regarded such changes as occurring gradually without necessarily being continuous. Moreover, he was well aware that they could take place more rapidly in some parts of the world than in others.

Since the mid-nineteenth century, geology and palaeontology have made great advances and, today, our knowledge of the broad evolutionary outlines of the main plant and animal groups is reasonably complete. Furthermore, as illustrated in Fig. 2.4, we now have a fairly precise idea of their duration throughout geological time. Patterson [2.4] has used such information to draw up a genealogical tree of the vertebrates showing their origins and relationships (Fig. 2.5). It will be seen

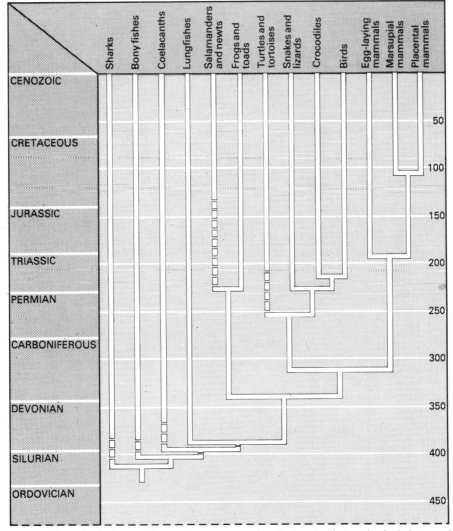

**Fig. 2.5** Genealogical tree of vertebrate groups in relation to the geological time-scale. Solid lines denote relationships known from fossils; dotted lines indicate gaps in the fossil record

that most groups are linked by solid lines indicating the existence of a fossil record. The gaps (represented by dotted lines) cover periods when the organisms are thought to have existed but where no fossil confirmation is at present available. This could be explained on the grounds of poor conditions for fossilisation, misidentification of existing fossils, or misconceptions regarding the particular portion of the ancestral tree. Incidentally, the existence of such gaps has also been cited by creationists as providing evidence against the theory of evolution (see p. 143).

## Missing links and living fossils

When one line of evolution diverges from another, we might expect to find intermediate forms possessing characteristics midway between ancestors and descendants. However, we must remember that in looking for such transitional stages, we are dealing with vast periods of time – millions of years – and genealogical trees which are usually far from linear and subject to considerable branching. But on rare occasions, fossil remains qualifying in varying degree as 'missing links' have been unearthed and these have played an important part in the unravelling of evolutionary relationships.

Reference to Fig. 2.5 will show that a solid line joins the birds and reptiles indicating a degree of fossil continuity. A vital link joining the two is the remarkable fossil bird possessing reptilian features, *Archaeopteryx lithographica* (Fig. 2.6), first found in 1845 in the limestone deposits of Solnhafen, Bavaria. The animals lived about 140 million years ago in the late Jurassic. At this time the dominant species over the Earth's surface were reptiles, so the development of powered flight represented an evolutionary step of major importance. Since the original discovery, several other specimens have been obtained and studied in great detail. From this it is clear that the animal was indeed a 'missing link' between reptiles and birds. Some of the characteristics of the two groups and the extent of their occurrence in *Archaeopteryx* are summarised in Table 2.4.

Confusion sometimes arises over the difference between the primitive birds such as *Archaeopteryx* and the pterodactyls (pterosaurs). These were essentially winged reptiles which acquired the power of gliding flight about the same time as

**Table 2.4**   Some characteristics of reptiles and birds

| Structure | Reptilian features | Avian features |
|---|---|---|
| Tail | Long, with free vertebrae at tip [A] | Vertebrae fused at tip, forming pygostyle |
| Sacrum | 6 vertebrae [A] | 11–23 vertebrae |
| Digits | None opposable | First digit (big toe) of hindlimb opposable [A] |
| Teeth | Present [A] | Absent in modern birds |
| Claws | Present on three digits of fore-limb [A] | None on wings of modern birds |
| Feathers | Absent | Present [A] |

[A], a feature of *Archaeopteryx*

(a)

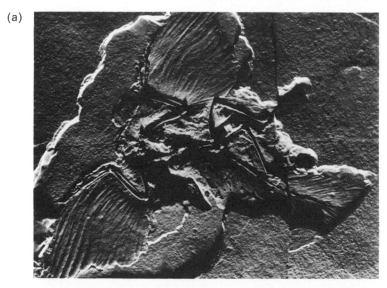

(b)

**Fig. 2.6** The oldest known bird, *Archaeopteryx lithographica*: (a) fossil, (b) reconstruction

birds, but in quite a different way. Thus in birds, the five-fingered (pentadactyl) forelimb is much reduced; the thumb is only a vestige and the wing is carried on the second and third digits. In pterodactyls, the wing membrane (which possessed no feathers) was stretched on a greatly extended little finger. The thumb was absent and the other three fingers were modified as claws (Fig. 2.7).

One of the principal problems of tracing evolutionary relationships from fossil evidence is that the preserved remains of dead plants and animals are seldom complete. The palaeontologist is then faced with a kind of jigsaw puzzle, fitting the various pieces together from a number of different finds and then reconstructing the organism as it probably existed when alive. The evidence on which such reconstructions are based is usually deficient in some respects, so the appearance of the final product always depends upon a few inspired guesses. On rare occasions, a species thought to be extinct turns out unexpectedly to have survived in a habitat which is either inaccessible or, for some reason, has been overlooked. Such an event occurred in December 1938, when a primitive fish known as a coelacanth was caught off the coast of East London in South Africa. The animal belonged to the Crossopterygii, a group of fishes with fringed fins. These are well

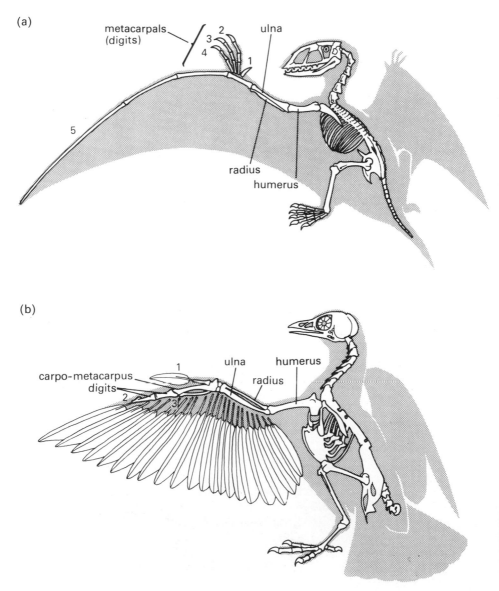

**Fig. 2.7** Modifications of the pentadactyl limb in (a) a pterodactyl, (b) a modern bird

known as fossils from the Devonian, about 230 million years ago, and represented a major evolutionary step in fish evolution. Thus, they had bony jaws, overlapping scales, and a skeleton which was, at least partly, of bone. Of particular significance were the fins which were like squat paddles with an internal skeleton. There seems little doubt that it was from ancestors such as this that the first land animals (amphibians) evolved, the paddles acting as primitive limbs. The fossil record of coelacanths terminates about 70 million years ago, since when, it had been assumed that they were extinct. Hence the excitement at the discovery of a 'living fossil'.

The first coelacanth captured was a large fish 1.6 m (5 ft) long and weighing 58 kg (127 lb); others caught subsequently have been about the same size. It was named *Latimeria chalumnae* after Miss M. Courtenay-Latimer, curator of the East London Museum, who was the first person to appreciate its significance as something out of the ordinary (Fig. 2.8). The story of the discovery and its sequel

(a)

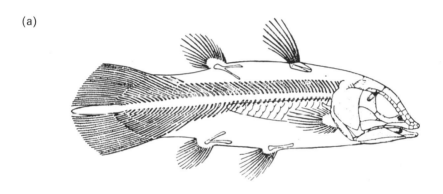

(b)

**Fig. 2.8**  A 'living fossil' – the coelacanth: (a) fossil reconstruction, (b) a specimen, *Latimeria*

have been well told and provide a fascinating account of detection and intrigue [2.5]. Unlike *Archaeopteryx*, *Latimeria* cannot be regarded as a 'missing link', although it provides valuable supplementary evidence to that already obtained from fossils as to what the ancestors of land vertebrates must have been like. In retrospect, perhaps one of the most interesting aspects of this extraordinary find is that it should have occurred at all, and that an animal of such characteristic appearance and biological significance should have remained undetected for so long. Its capture alive provided a unique opportunity of gauging the accuracy of previous reconstructions based on fossil evidence. In the event, these proved to be remarkably life-like (Fig. 2.8).

## Fossils and gradual change

Darwin's idea of gradual evolutionary change (sometimes referred to as **gradualism**) is well illustrated by the fossil history of several groups of organisms. One of the best documented is that of the horse whose ancestral history was first studied as long ago as 1879. Fossil remains of horses have been found almost the world over, some of the earliest being discovered in Suffolk, England. The Eocene form of horse, which existed some 60 million years ago, was little larger than a fox terrier and about 0.4 m high (Fig. 2.9), with four toes in front and three behind (Fig. 2.10). When first described, it seemed so unhorse-like that it was named *Hyracotherium* (later *Eohippus*). Evidently, it was an animal of forests and the structure of its teeth suggests that it was a browser living on plant leaves and stems. Subsequent climatic changes resulted in a gradual drying of the land, and these were accompanied by alterations in the horse's anatomy and behaviour from being a pad-footed forest dweller to a more rapidly moving spring-footed inhabitant of the plains. Large carnivores were already in existence, so increased

**Fig. 2.9**  *Hyracotherium* (*Eohippus*), the dawn horse from the Eocene. It had four toes on the front feet and three on the hind feet

speed was no doubt an advantage for survival. In the Oligocene we find *Mesohippus*, a larger animal of about 0.6 m height with three front toes and a longer neck – no doubt an adaptation for grazing. By the mid-Miocene horses had become spring-footed with the two side toes still present but reduced in size, while by the late Miocene the side toes of *Pliohippus* (1.2 m high) had disappeared altogether. The first modern horses (*Equus*) appear at the end of the Pliocene and were widespread from the early Pleistocene onwards.

The stages in the evolution of horses thus correlate well with variations in ecological conditions, which the fossil record tells us took place during a period of around 70 million years from the Eocene until the beginning of the Pleistocene. Bodily changes involved not only a great increase in size (from 0.4 to 1.6 m high) and adaptations of the limbs, but also modifications of the skull, brain, and teeth. But in using the example of the horse it would be wrong to gain the impression that the evolutionary series represents a tidy linear sequence. The amount of good fossil evidence available is enormous and from this it is clear that there were numerous side branches from the main line of development, many of which became extinct. Thus, *Hyracotherium* (*Eohippus*) was present during Eocene times in both the Old and New Worlds. By the Oligocene the Old World stock was extinct and no fossils are found until the Miocene when they reappear, presumably as a result of migration from elsewhere and subsequent recolonisation.

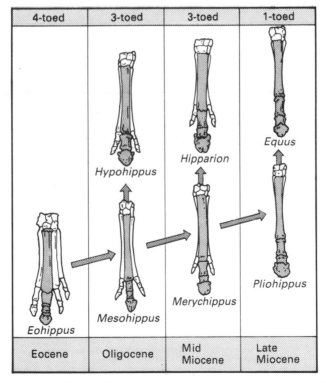

**Fig. 2.10** Evolution of the forelimb in the horse over a period of about 60 million years 1982)

## Fossils and interrupted change

The evolution of the horse over a period of around 70 million years presents a picture of a group of organisms faced with fundamental changes in the environment and gradually adapting to them. But, as Stanley [2.6] has pointed out, if we look more closely at the relative times when the various genera made their appearance, we find considerable variations in the gaps between them. For a detailed scrutiny of such time differences it is necessary to have a more complete fossil record than there is at present. However, such evidence as exists suggests that although the later evolution of the horse may have been fairly continuous, conforming to a gradualistic pattern, the early stages were discontinuous. Thus, two species of the so-called 'dawn horse', *Hyracotherium*, existed in the woodlands of North America for at least 3–4 million years without apparently undergoing any appreciable change in appearance over a span of approximately a million generations.

Seen in this light, the history of the horse poses a fundamental question. Did its evolution proceed in a gradual manner as envisaged by Darwin and traditionally believed, or did it occur as a series of discontinuous (punctuated) spurts? Or, put more generally, should the evolution of the horse be explained in terms of a **gradualistic** or **punctuational** process – or both?

Among vertebrates, which provide some of the richest and most reliable fossil evidence, numerous other examples comparable with the early horses lend support to the idea of at least some punctuational evolution. Thus, the mammoth, *Mammothus* sp. (Fig. 2.11), a hairy elephant-like creature with huge tusks, ranged over Europe for about $3\frac{1}{2}$ million years from the time of its first appearance to its extinction. There seems some doubt as to whether the three different species that may have existed were, in fact, distinct and to what extent they represented a single line of evolution. The fact remains that over this period no fundamental changes in structure or other adaptive modifications seem to have occurred.

**Fig. 2.11**   Reconstruction of the European mammoth, *Mammothus* sp.

One of the most striking examples of discontinuity in evolution has been described by Williamson [2.7] and concerns invertebrates. Both gastropod and bivalve molluscs fossilise well because their shells, once petrified, retain their original condition usually in a good state of preservation. A recently discovered series of such shells from Pliocene–Pleistocene deposits in the Turkana Basin of northern Kenya has provided one of the most complete fossil sequences ever studied. A portion of these results covering aquatic gastropods is summarised in Fig. 2.12. The general pattern of their evolution appears to conform to a punctuated rather than a gradual pattern. Thus, in the thirteen sequences which were sufficiently complete for detailed analysis, long periods of stability in structure and size covering millions of years, were interrupted by short bursts of rapid change of a duration of 5000–50 000 years. The newly evolved forms then persisted unchanged for a further extensive period, many eventually becoming

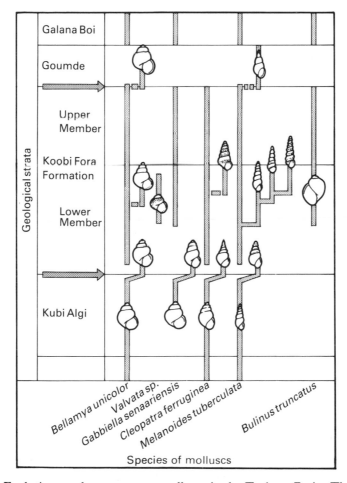

**Fig. 2.12** Evolutionary changes among molluscs in the Turkana Basin. The heavy arrows show points in the fossil sequence where sudden evolutionary changes took place simultaneously among a number of different species of gastropods (after Williamson 1982)

extinct. These in turn were replaced by types resembling closely the ancestral pattern, some of which still persist today. In evaluating these results, two particular features should be borne in mind:

(i) Relative to the period of geological history involved, the appearance of forms intermediate between ancestors and descendants occupies only a tiny proportion of the total time.

(ii) Within the different genera of molluscs, the periods of transition between one form and another roughly coincide, suggesting that the factors that prompted change operated similarly on them all.

Discontinuities of the kind outlined above have, in fact, been known for some time, although they have never been analysed in such detail. Thus, as we saw earlier (p. 21), Cuvier was faced with similar problems in his studies of Tertiary fossils in the Paris Basin. He interpreted the presence of fossils in some strata and their absence from others as being due to a series of catastrophes (p. 22). How do we account for such situations today? Do gradual and punctuated patterns of evolution represent separate entities or are they merely different parts of a single evolutionary spectrum? To what extent are our present ideas on evolutionary theory capable of accommodating and explaining the current evidence from geology? These are important questions that need to be resolved. But before tackling them (see Chapter 8), we must be clear about the biologist's viewpoint on evolution and, in particular, the directions in which evolutionary thinking has moved since the time of Darwin.

# Viewpoints of geology and biology

One of the strongest supporters of a punctuational explanation of evolution has observed that those who adhere to the idea 'admit that large, fully established species evolve but [we] see these species as typically changing very little before becoming extinct. The lineages that they form may nonetheless sometimes change enough that the oldest and the youngest of their fossil populations cannot be accommodated comfortably within a single species' [2.6].

This statement serves to highlight the different approaches of the geologist and biologist. If we regard the whole of life past and present as a tree, the main concern of biology is only with the tips of the branches that represent the present. The remainder extends below and constitutes the past, and is therefore subject to the historical limitations that we have encountered earlier. Nowhere are these limitations more apparent than in the concept of **species**. In deciding whether two organisms belong to the same species, the biologist uses such criteria as similarities in structure, ecology, distribution, and the ability to interbreed and produce fertile offspring. The geologist has to approach the species problem in a more arbitrary manner by chopping the tree of life and its branches into numerous apparently intergrading parts. The limits of each part may have to be defined only in statistical terms using such parameters as shape, size, and superficial detail. Such groups are known as **chronospecies**. Stanley [2.6] has defined a chronospecies as 'a segment of a lineage judged to encompass little enough evolution that the individuals within it can be assigned a single species name'. Biologists and geologists thus differ fundamentally in their concept of species; this is an important consideration when attempting to account for the evolution of life at different levels of complexity, both now and in the past.

# Summary

1 The effectiveness of fossilisation in preserving the remains of dead plants and animals varies with circumstances. Sometimes the fossil record is reasonably complete but more often it is not.

2 In the early nineteenth century, Cuvier interpreted fossil discontinuity as evidence of extinction due to a series of catastrophes, the latest being the biblical flood. His idea of progressionism stated that each successive group of organisms was more advanced than the one before it.

3 Lyell's principle of uniformity maintained that the same agents were responsible the world over for laying down geological strata, both now and in the past. There was therefore no need to invoke catastrophes to explain fossil discontinuities.

4 Darwin's geological observations on land elevation in South America added an important quantitative dimension to the findings of Lyell.

5 In *The Origin of Species*, Darwin stressed the inadequacies of the fossil record. Fossils contributed little to his theory accounting for change but they provided important evidence that change had occurred in the past.

6 Since the nineteenth century, great advances have been made in our knowledge of the fossil record. Today, the main outlines of the ancestry of plants and animals are reasonably complete.

7 The Quaternary period in Britain was of particular importance in establishing our present-day flora and fauna.

8 Fossil links between related ancestral groups of organisms such as *Archaeopteryx*, are important in establishing evolutionary stages in the modification of different organs for new purposes, as in the evolution of birds from reptiles.

9 'Living fossils', such as the coelacanth, *Latimeria*, can be valuable both as a check on previous knowledge of relationships between different groups of organisms and as a means of verifying reconstructions based on fossil evidence.

10 Darwin believed that the evolution of life was a gradual process. Many groups of organisms appear to exhibit a gradualistic pattern of change over time.

11 There is evidence, however, that evolution also proceeds discontinuously, being rapid at one juncture and slower at another. This punctuational pattern of change could be more widespread than is generally realised.

12 When attempting to resolve problems such as the relative significance of the gradual and punctuated patterns of evolution, it is important to appreciate the divergence between geological and biological viewpoints on such concepts as species.

# Topics for discussion

1 There are still numerous gaps in the fossil record. What is the likelihood of finding fossil remains of a plant or animal that bear no resemblance to any known form either living or extinct?

2 What justification is there for saying that while Lyell was the first uniformitarian in geology, Darwin was the first uniformitarian in biology?

3 The search for missing links in the fossil records of some plant and animal groups has been a mixed blessing in helping to unravel their evolution. What are the advantages and disadvantages of such an approach?

4   It has been suggested that to try to distinguish between gradualistic and punctuational models of evolutionary change is to draw a false comparison. The difference between the two is merely one of degree. To what extent is this a valid argument?

5   Suppose that fossils were able to provide information on the soft parts of extinct organisms and on the ecological conditions in which the organisms existed, how would such knowledge influence our concept of species?

# References

2.1   Hamilton, W. R., Wooley, A. R., and Bishop, A. C. (1974) *Minerals, Rocks and Fossils*. Hamlyn

2.2   de Beer, Sir G. (1963) *Charles Darwin*. Thomas Nelson

2.3   Wilson, E. O. *et al*. (1973) *Life on Earth*. Sinauer Associates, Sunderland, Mass.

2.4   Patterson, C. (1978) *Evolution*. Routledge and Kegan Paul

2.5   Smith, J. L. B. (1956) *Old Fourlegs*. Longmans

2.6   Stanley, S. M. (1981) *The New Evolutionary Timetable*. Basic Books, New York

2.7   Williamson, P. G. (1982) 'Palaeontological documentation of speciation in Cenozoic molluscs from Turkana Basin', in *Evolution Now: a Century After Darwin* (ed. J. Maynard-Smith). Nature/Macmillan

# 3

# Darwin's explanation of change

During his voyage in HMS *Beagle*, Darwin (Fig. 3.1) accumulated not only geological and fossil evidence but also much biological information to support his view that evolution had occurred in the past and was still proceeding. At the various ports of call along the east coast of South America, he collected and observed many kinds of local plants and animals in their native habitats. He noticed that some species varied appreciably from one locality to another; the closer they were together the less was divergence apparent. Thus, in Argentina and Uruguay he saw great numbers of the ostrich-like rhea, *Rhea americana*, while further south in Patagonia he encountered a dark-coloured, smaller species later called *Rhea darwinii*. Separation of the two appeared to be due more to the geographical distance between them than to any physical barrier, such as a mountain range or area of water. Local variations were also seen among the characteristic rodents such as the agoutis, coypus, and capybaras. But the most striking feature common to them all was that their structure and behaviour conformed to a pattern that was essentially South American, and quite distinct from their North American counterparts occupying similar ecological niches.

## The Galápagos Islands as an outdoor laboratory

By far the most powerful biological evidence for change acquired by Darwin was during a 5-week visit to the Galápagos Islands in 1835. The Galápagos archipelago is a group of volcanic islands lying on the Equator about 1000 km (600 ml) to the west of Ecuador on the South American mainland (Fig. 3.2). The islands vary greatly in size from mere rocks to the length of Albemarle, about 120 km (75 ml) with a volcano 1650 m (5400 ft) high. The vegetation of the small islands is sparse and windswept, but the larger islands have a diverse flora that includes rain forests at an altitude of between 220 m (750 ft) and 440 m (1500 ft). Darwin soon realised that here was the equivalent of an outdoor biological laboratory, isolated from the mainland by a vast stretch of the Pacific Ocean, where he could study the ways in which different species had diversified and adjusted to a great variety of ecological habitats.

But the most striking feature of the Galápagos flora and fauna, which influenced Darwin profoundly, was the fact that although many species were clearly peculiar to the islands, all showed an unmistakable resemblance to those occurring on the South American mainland, from which they had presumably been derived in the remote past. Isolation, inherent variation, and competition for the available ecological niches must have been responsible for the patterns of **adaptive radiation** that were so clearly evident. These were particularly striking among the reptiles and birds.

The largest indigenous land animals on the Galápagos are the tortoises, all of which belong to a single species, *Geochelone elephantopus* (indeed, the word Galápagos is the ancient Spanish for tortoise). These huge creatures reach a length of about 2 m (6 ft) and a weight of some 270 kg (600 lb). On visiting the islands,

**Fig. 3.1** Charles Darwin in 1840, four years after returning from his voyage in HMS *Beagle*

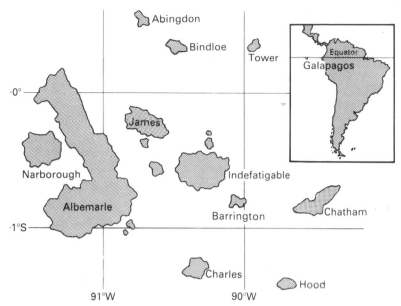

**Fig. 3.2** The Galápagos Islands (smaller islands omitted). The English names are those used in Darwin's time

Darwin was informed by the vice-governor that the inhabitants could tell to which island a particular tortoise belonged by its peculiar characteristics. This claim was later substantiated and some fifteen subspecies are now recognised, differing from one another in such characteristics as size, shape, colour, shell thickness, and length of neck and legs. The tortoises can be divided into two groups (Fig. 3.3). One has a domed shell and a short neck (Fig. 3.3(a)), and lives in grassy areas where it grazes on low-growing vegetation. The other has long legs and a shell that is raised in front with a wide (flared) gap giving the neck considerable flexibility (Fig. 3.3(b)). This latter group is found in arid habitats where the predominant vegetation consists of small shrubs on which the animals browse. The main food of the tortoises is the prickly pear cactus, *Opuntia* spp., and on islands where the animals are absent, these plants grow in a characteristic dwarf and spreading manner, being covered with soft spines. But where *Opuntia* is subjected to the grazing pressure of tortoises, the growth pattern is erect and tree-like, and the spines are stiff and protective. Evidently, the plants and animals have evolved together – an example of **coevolution**.

### Birds of the Galápagos

More than any other group of animals, the birds of the Galápagos Islands provided for Darwin an extraordinary museum of adaptation. They must have shattered any lingering doubts that he may still have harboured regarding the likelihood of evolution. Some of the more extreme examples of bird species are widely distributed on many of the islands and include a flightless cormorant with only vestigial wings, which swims to catch its food of fish but is otherwise terrestrial. Again, among maritime species, the cormorant-like blue-footed booby (*Sula nebouxii*) nests in colonies on the rocks. However, one species, the red-

(a)

(b)

**Fig. 3.3** Galápagos giant tortoises, *Geochelone elephantopus*: (a) short-necked form with domed shell, (b) long-necked variety with flared shell

footed booby (*Sula sula*), has acquired the habit of nesting in trees, thus avoiding competition for territories with its near relatives. The capacity for prising insects out of crevices in volcanic rock and tough vegetation has also been well developed. Thus, the Galápagos mockingbird (*Nesominus trifasciatus*) resembles its North American counterpart but the beak is longer and more slender enabling it to pry into lava cracks. Most remarkable of all is the woodpecker finch (*Camarhynchus pallidus*) (Fig. 3.4), which has acquired the habit of employing cactus spines or small twigs to prise insects out of crevices in trees. There are no woodpeckers in the Galápagos, and as a result of this remarkable adaptation the finch has succeeded in occupying the woodpecker's usual ecological niche.

**Fig. 3.4** The woodpecker finch (*Camarhynchus pallidus*). The bird has acquired the habit of using a cactus spine or twig to prise insects out of crevices

## Finches

In his diary, Darwin refers to 'a most singular group of finches'. These consist of some thirteen species with dull plumage, resembling sparrows in their general appearance.

The ground finches (*Geospiza* spp.) live near the coast and, as their name implies, feed mainly on the ground, predominantly on seeds but also on some insects. Their beaks are relatively large and heavy; an effective adaptation for cracking the shells of fruits and seeds (Fig. 3.5(a) and (b)). The tree finches (*Camarhynchus* spp.) live in the forest on a mixed diet of insects and a fair proportion of plant material; their beaks exhibit a somewhat intermediate condition (Fig. 3.5(c)). The warbler finches (*Certhidia* spp.) feed exclusively on insects, their habits resembling somewhat those of flycatchers; their beaks are slender and typical of an insectivore (Fig. 3.5(d)). Unfortunately, we know nothing of the range of species that originally colonised the islands from the South American mainland. But all the evidence suggests that from this ancestral stock, a number of divergent lines developed which gradually became adapted to a variety of different ecological conditions – a beautiful example of **adaptive radiation**.

Variation among the Galápagos finches can be considered at two levels: major variants such as beak structure, enabling the animals to take over specific ecological niches; and minor variations peculiar to each island. Thus Lack [3.1] studied such characteristics as beak size and wing dimensions, and showed that such second-order differences were also a feature of the various islands, as is illustrated by the ground finch *Geospiza fuliginosa* (Fig. 3.6).

Using the analogy of crossbill migrations in Europe, Harper [3.2] has argued that the situation of the Galápagos finches could equally well be explained by

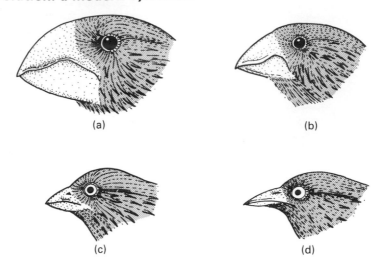

**Fig. 3.5**  Heads of four species of Galápagos finches showing variation in beak structure correlated with diet: (a) *Geospiza magnirostris*, food large seeds; (b) *G. fortis*, food smaller seeds; (c) *Camarhynchus parvulus*, insectivore; (d) *Certhidea olivacea*, insectivore (small insects)

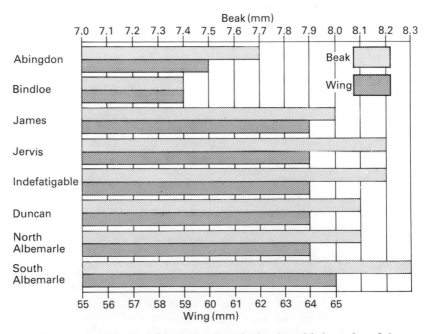

**Fig. 3.6**  Mean measurements of beak depth and wing breadth in males of the ground finch, *Geospiza fuliginosa*, on different islands of the Galápagos; average sample size about 45 (data from Lack 1947)

postulating that the various species reached the islands as periodic irruptions of mainland populations, displaying much the same characteristics as they do today. On this assumption, evolution in the Galápagos themselves would have been confined to second-order variations only. He supports this contention by calculating the target size of each island as it would appear to a migrating bird, and shows that the larger the area of land, the greater the number of species it supports. However, following Darwin's line of reasoning, it could just as well be argued that the larger the area occupied by a species, the more diversified it is likely to become, and hence the greater the opportunity for evolving different forms adapted to specific circumstances.

But no matter how we interpret the origins of the Galápagos finches, the fact remains that at some time and somewhere adaptive radiation resulted in the diversity of species that Darwin observed. Such observations provided him with one of the essential links in his theory of evolutionary change, which he was later to expound with such effect in *The Origin of Species*.

## Darwin and Malthus

Two years after his return to England in 1836, Darwin happened to read 'for amusement' the *Essay on Population* by the Revd Thomas Robert Malthus which had been published in 1798. In this Malthus claimed to show that any attempt by man to produce a more perfect society was doomed to failure. This gloomy view was based on the assumption that although populations increase exponentially (i.e. in a geometrical ratio), food resources increase only in a linear manner (i.e. in an arithmetical ratio). For the well-to-do, this need not be a serious consideration, but among the poor, 'the superior power of population cannot be checked without producing misery or vice'. It is sometimes claimed that Darwin deduced from Malthus's essay the conclusion that in the struggle for survival that must ensue from limited resources, those best adapted to their environment will, on average, survive and breed at the expense of the less well endowed, who will tend to be eliminated. This was the principle of **natural selection**. But as de Beer [3.3] has pointed out, reference to Darwin's notebooks on the transmutation of species shows clearly that he had already formulated this idea while he was still in South America. Malthus's gift to Darwin was the view (incidentally erroneous!) that human populations increased exponentially. If this were applied to wild populations of plants and animals as well, it followed that natural selection was the inevitable consequence.

## Darwin and Wallace

The publication by Darwin of his explanation of evolutionary change took place in a series of stages. Much concerned about his health he realised that provision must be made for the publication of his life's work in the event of his death before the completion of the final version. Accordingly, he prepared an outline summary, the *Sketch*, which was completed in 1842. This was followed by an expanded version, the *Essay*, of 1844. Both remained unpublished until the Darwin–Wallace centenary in 1958 [3.4]. Meanwhile, one of the most remarkable coincidences in the history of science occurred when another English naturalist, A. R. Wallace (1823–1913), then exploring the Malay Archipelago, wrote to Darwin proposing a mechanism of evolution by natural selection precisely similar to his own. A situation requiring considerable delicacy and tact was admirably handled by Lyell and the botanist J. D. Hooker, and resulted in a joint paper by Darwin and Wallace, 'On the Tendency of Species to Form Varieties; and on the Perpetuation of Varieties and Species by Natural Means of Selection', also two separate ones which were read before the Linnean Society of London in 1858. Curiously, the papers made relatively little impact. However, Darwin now realised that publication of his findings that he had gradually assembled in the *Sketch* and *Essay* was a matter of urgency, and in 1859 he produced his famous work *The Origin of Species by Means of Natural Selection or the Preservation of Favoured Races in the Struggle for Life* (generally shortened to *The Origin of Species*). The existing state of opinion on the subject of evolution, together with a widespread interest in natural history no doubt account for the enthusiasm with which *The Origin* was received by the educated public of the Victorian age. The first edition of 1250 copies was sold the day it was published and 13 years later the book had already reached its sixth edition.

## Is Darwin's explanation a theory?

Today, we regard Darwin and *The Origin of Species* as the fountain source from which modern concepts of the mechanism of evolution are derived. But we must remember that his prime concern was not to explain evolution, but how new species are formed. Nowhere in *The Origin of Species* is there any reference to the word 'evolution'.

The question is sometimes asked whether Darwin's views on the origin of species qualify as a theory. In essence, a scientific theory is a statement of one or more principles, usually derived from a causal relationship between sets of facts. An important feature of all theories is that they must be susceptible of scientific test which, as the philosopher Karl Popper pointed out, can disprove them but cannot establish their final validity. The principles of natural selection can be summarised as four propositions (inductions), each of which is theoretically testable:

  (i) In any generation, not all individuals of a species succeed in reproducing.
 (ii) Members of a species are not identical; they exhibit individual variation.
(iii) Apart from variation due to environmental differences, the majority is inherited and transmitted from parents to offspring.
(iv) Reproductive success is not random. It is associated with inherited characteristics, some of which will be more beneficial in particular circumstances than others.

From the four propositions, the existence of natural selection inevitably follows.

An essential feature of Darwin's concept of natural selection was that it not only postulated differential survival of those individuals best adjusted to a particular environment, but also reproductive superiority in producing a greater number of offspring. Selective advantage was closely coupled with enhanced breeding success. To what extent did these propositions lend themselves in practice to scientific test?

## Some problems facing Darwin

Darwin was acutely aware of the defects of his arguments put forward in *The Origin of Species*. Much of the book is concerned with a mass of evidence accumulated in support of his ideas, an emphasis on the gaps in existing knowledge, and doubts associated with them. These were particularly evident in two areas of vital importance – **variation** and **natural selection**.

As a result of his many observations on varieties in wild populations of animals and plants, Darwin was in no doubt that much of this variation was inherited. Moreover, his detailed knowledge of the achievements of domestication served to reinforce his appreciation of the value of variation as a potentiality for selective change. But he knew nothing of the mechanism of inheritance and no coherent theory was forthcoming during his lifetime. A second problem concerned the action of natural selection, which Darwin had deduced as an inevitable outcome of the struggle within species for survival. In his time, evidence for its existence was only anecdotal, and it was not until the 1920s that the study of selection in action was placed on an experimental basis.

## The origin of variation

In the late eighteenth century, a good deal of research had been carried out (mainly in Germany) on the effects of crossing different varieties of plants. The conclusion invariably reached was that offspring tended to exhibit characteristics intermediate between those of their parents. This had led to the idea of **blending inheritance**, which implied that at each successive generation the fund of inherited variability tended to be reduced, with an inevitable trend towards increasing uniformity. Darwin was aware of the basic limitations of this process, and realised that some mechanism must exist in the bodies of organisms whereby inherited material could be passed on from one generation to the next. He postulated a theory of **pangenesis** involving hereditary 'gemmules' that were transmitted at sexual reproduction, and this played some part in influencing the scientific outlook of his time. But he was at a disadvantage in having no clear idea of the distinction between **germ cells**, concerned exclusively with reproduction, and **somatic (body) cells**, with no reproductive function. The continuity of the germ plasm, as it came to be called, from one generation to the next was not finally established until the work of A. Weismann in 1892, some 33 years after the publication of *The Origin of Species*. The outcome of this was at once far reaching and, among other things, evoked the famous comment by Samuel Butler that 'a hen is only an egg's way of producing another egg'.

Darwin's observations in the field had convinced him that all living organisms periodically undergo spontaneous changes in structure and appearance, many of which are inherited. He considered these to be of two kinds:

(i) Large variations that were easily recognisable. These he called 'sports'. Their essential feature was their uniqueness, in that they could not be regarded as part of a graded series. Today we would call this **discontinuous variation** (see p. 58).

(ii) 'Fluctuating variation' that was less obvious and extreme, and often formed part of a graded series, such as human hair colour. Its description is near to what we would now call **continuous variation** (see p. 58). It was this that provided both the fund of variance on which natural selection must act if evolution were to come about and the reservoir from which the losses due to blending inheritance could be made good.

It is doubtful if Darwin ever read the work of Gregor Mendel, first published in

1866, in which he demonstrated the existence of **particulate inheritance** (page 56) through his pea experiments. Even if he had been aware of these findings, it is unlikely that he would have appreciated their significance for the origin of species. In the mid-nineteenth century, genetic studies were dominated by Francis Galton and his associates, who were largely concerned with investigating the transmission of such identifiable characteristics in families as genius, and the correlation of one set of variables with another. Such studies are now known as **biometry**. As would be expected, the characters chosen for study mostly exhibited continuous variation, and this explains to some extent the emphasis placed by Darwin on this aspect of variability. No coherent explanation of genetic phenomena ever emerged in his time and it was not until 1900, when the work of Mendel was rediscovered and its significance appreciated, that one of the essential links of evolutionary theory was finally set in place.

Another observation made by Darwin concerned the range of variation existing among domesticated species, which seemed to exceed by far that occurring under wild conditions. He explained this, at least in part, on grounds of environmental influences such as an abundance of food, a distinctly Lamarckian approach (see p. 144). This only serves to highlight the diversity of rival views on the nature of inheritance that existed in the mid-nineteenth century.

### Evidence for natural selection

Darwin equated natural selection with the 'survival of the fittest', a term coined by Herbert Spencer after the publication of *The Origin of Species* and unfortunately included in later editions (the title of Chapter 4) where it tended to gloss over his real meaning. Darwin's criterion of fitness was the degree of adaptation judged by the ability of an individual to survive in a particular set of circumstances (i.e. its **viability**). However, another important measure of fitness is the number of progeny produced by a well-adapted individual relative to those of a less well-adapted one (i.e. its **fertility**). As Edwards [3.5] has stressed, an organism that survives but makes no contribution through reproduction to the next generation can hardly be said to be 'fit' in a Darwinian sense.

From the propositions outlined earlier (p. 48) Darwin deduced that natural selection must occur. Although he was well aware of its outcomes in populations such as the Galápagos finches, he had no evidence other than of an anecdotal kind, to explain *how* it came about, such as the **selective** agents involved or the nature of their action (i.e. **selective advantages** and **disadvantages**). The development of an objective and experimental approach had to await the beginning of ecological genetics in the 1920s.

### Artificial selection and domestication

The nearest Darwin got to studying natural selection in action was in his close association with horticulturists and animal breeders who practised artificial selection in the production of new varieties. Thus, regarding domestic dogs, he asks, who would believe that animals resembling the greyhound, bloodhound, bulldog, and pugdog, all so unlike any known wild species, could ever have existed in a state of nature? As we now realise, a breed such as a fox terrier can alter

appreciably in appearance as a result of intense selection of particular characteristics even within a human lifetime.

An example quoted in *The Origin of Species* is that of domesticated pigeons,
which, as a result of human selection, have radiated into some extraordinary
aberrations (Fig. 3.7). It is thought that the common ancestor of the 300 or so
breeds existing now was the rock dove, *Columba livia* (Fig. 3.7(a)), whose
distribution today is restricted to southern Europe, northern Britain, and Ireland.
One of the commonest domestic strains in Britain, the fantail, may have originated
in Asia. It is usually white, but can be other colours such as black and yellow (Fig.
3.7(b)). The pouter (Fig. 3.7(c)) is thought to have been brought to Britain some
five centuries ago by Dutch traders returning from India. It occurs in several
different colours and is characterised by its huge barrel-like 'chest' and long,
slender legs. The jacobin (Fig. 3.7(d)) is so-called on account of its curious monk-
like hood of feathers, whose size and degree of elaboration are much prized by
breeders. Other oddities of human selection have extended to behaviour patterns,
such as the bizarre aerobatics of the tumbler pigeon. In modern times, the power
and effectiveness of artificial selection have been put to practical use in virtually
every sphere of stock and crop production, ranging from sheep and cattle to fruit
and vegetables, and to ornamental plants such as roses, dahlias, and begonias.

**Fig. 3.7** The ancestry of some domestic strains of pigeons: (a) the rock dove, *Columba
livia* (thought to be the ancestral form), (b) the fantail, (c) the pouter, (d) the jacobin

## Darwinism in the nineteenth century

In his preface to the reprint of the sixth edition of *The Origin of Species*, de Beer [3.6] writes,

> There is a small number of great books which have changed the face of the earth. Such as the Bible, they have exerted their effects universally even if their tenets are not everywhere accepted. Others, like Newton's *Principia*, have revolutionised the state of thought and of material condition in which men live, whether they are aware of it or not. It is to this latter category that Darwin's *Origin of Species* belongs.

Throughout this chapter we have been concerned with the lines of evidence that led to the publication of this great book and the eventual acceptance of the set of principles it contained. These converge, as it were, from several different directions and, as a conclusion, it will be worth while summarising them in order to clarify the position of evolutionary thought at the end of the Darwinian era.

One of the greatest advances in nineteenth-century biology was the establishing, once and for all, that evolution had occurred in the past and that it was still taking place, thereby deposing the previously held doctrine of Divine Creation. Two fields of discovery were responsible for this radical change of outlook. One, the study of geology, had made great advances under the leadership of Sir Charles Lyell. As we saw in Chapter 2 (p. 22), the theory of uniformity transformed existing ideas about the age and construction of sedimentary rocks, and led to a significant increase in our knowledge of the fossil record. To this, Darwin himself made an appreciable contribution, as a result of his studies in South America. Further evidence for evolution came from Darwin's observations of living organisms, in particular, the role of **isolation** in the development of new species and varieties. His experiences with the vertebrates of South America and the finches of the Galápagos Islands were of particular significance in establishing the principle of adaptive radiation; it was this that finally disposed of the possibility that individual creation might have occurred.

Darwin realised that the raw material of evolution was **variation**, but here he was faced with a curious paradox. On the one hand it was clear that if evolution were to proceed, a continuous supply of new variants must be forthcoming. This he accounted for through small spontaneous changes (fluctuating variation) in the hereditary material; today we would call them **mutations**. But for beneficial variations to be preserved and transmitted, a mechanism of inheritance for carrying these traits uncontaminated was also necessary. The only system then accepted was **blending inheritance**, which resulted in a **reduction** in variation – the exact opposite of what Darwin required for his theory. He was therefore driven to postulating a high rate of sporting (**mutation rate**), which conflicted with all his observations of natural populations in the field. These had suggested that spontaneous inherited changes were, in fact, quite rare. The apparent contradiction was not resolved until the rediscovery of Mendel's work in 1900.

Lastly, there were problems posed by **natural selection**. Darwin had come to the conclusion that this must occur through a sequence of inductive reasoning based on the premise that rapid increases in numbers would outstrip available resources, resulting in a high mortality and differential survival of the best-adapted individuals. But apart from casual observations, he had no evidence in support of the action of natural selection in the wild. As was explained earlier (p. 50), this no doubt accounts for the importance he placed in *The Origin of Species* on evidence from artificial selection and domestication.

The nature of natural selection and selective agents, the magnitude of selective advantages and disadvantages, and rates of evolutionary change, are all issues that have gained prominence since the end of the nineteenth century as a result of the development of new mathematics and improved experimental techniques. Perhaps the most burning question of all is whether the mechanism of evolution, as we know it (sometimes referred to as neo-Darwinism [3.7]), is sufficient to account for the changes that are now known to have occurred over time in the plant and animal populations of the Earth, as a result of the greatly improved fossil record. This, and other kindred issues, will be considered further in the chapters that follow.

## Summary

1   The voyage in the *Beagle* proved a turning point in Darwin's thinking about evolution. In addition to geological evidence, he made numerous observations on local varieties, particularly South American vertebrates including the influence of isolation on their origin.

2   The Galápagos Islands provided Darwin with an outdoor laboratory for the study of evolution. In particular, the giant tortoises showed evidence of adaptive changes in response to different environments, also of coevolution with the plants on which they fed.

3   The Galápagos finches were of special interest having undergone adaptive radiation in adjusting to the requirements of a wide range of ecological conditions.

4   The influence of Malthus on Darwin's theory of the origin of species is often misconstrued. His principal contribution was to point out that whereas human populations might be expected to expand geometrically, food supplies are unlikely to increase in more than a linear manner. Darwin extended this idea to wild populations.

5   Darwin and Wallace conceived the idea of evolution through natural selection at the same time and produced joint papers for the Linnean Society. This novel approach to the origin of species made little impact on the scientific community.

6   The publication of *The Origin of Species* was preceded by the *Sketch* and *Essay*. Its appearance caused an overwhelming response and was greeted with enthusiasm in some quarters but by antagonism in others, particularly the Church.

7   Darwin's ideas on change consisted of four propositions from which the existence of natural selection inevitably followed. Unlike previous explanations of evolution, in theory, all were susceptible of scientific test.

8   Darwin realised that variation was the raw material of evolution. However, the mechanism of blending inheritance, to which he subscribed, tended to reduce variance rather than to increase it.

9   In equating natural selection with survival, Darwin stressed adaptability and the capacity to exist in different conditions. Fertility, as a component of selective advantage, was underemphasised.

10   Darwin was unaware of natural selection in action. The nearest he came to it was in his association with horticulturists and animal breeders. Hence the stress he laid on the power of domestication through artificial selection.

11   One of the greatest advances of nineteenth-century biology was the realisation that evolution had not only occurred in the past but was still taking place. This was due partly to the rise of geology at the hands of Lyell and others, and partly also to Darwin's observations of fossils and adaptive radiation among the living plants and animals of South America and the Galápagos Islands.

12   In order to offset the adverse effects of blending inheritance on variability, Darwin was compelled, against his better judgement, to postulate a high rate of sporting (mutation).

13   The origins of variation, the mechanism of inheritance, and the mode of action of natural selection, remained unresolved at the end of the nineteenth century.

## Topics for discussion

1   In Victorian times, the idea that man had an ape-like ancestry was widely considered repugnant. The same is not true today. What factors have brought about this change of attitude?

2   Darwin demonstrated convincingly the power of artificial selection in changing the characteristics of species such as pigeons. To what extent, if at all, should we apply this principle to the eradication of deleterious inherited diseases in man? What are the scientific and ethical problems involved?

3   In spite of his views on the action of natural selection, Darwin sometimes resorted to Lamarckian ideas to explain the adaptations he observed. How was this possible when the two approaches to the explanation of change are so different?

4   Darwin set sail in HMS *Beagle* in 1831, a fundamentalist (believing in the literal truth of Genesis) and returned 5 years later an evolutionist. What influences caused him to change his mind?

5   To what extent can Darwin's explanation of the origin of species be regarded as a scientific theory?

6   Suppose Darwin had read and appreciated the significance of Mendel's paper on heredity in peas, what difference could this have made to his theory of the origin of species?

7   Weismann's theory of the continuity of the germ cells provides an essential component of any theory of organic evolution. Why is this so?

8   Why is it that domesticated varieties of plants or animals produced by the process of artificial selection would be unlikely to survive under natural conditions, even though many of their characteristics are better developed than their wild counterparts?

9   What justification is there for the claim that *The Origin of Species* is one of 'a small number of great books which have changed the face of the earth'? (de Beer 1956).

# References

3.1   Lack, D. (1947) *Darwin's Finches*. Cambridge University Press
3.2   Harper, G. H. (1980) 'Speciation or irruption: the significance of the Darwin finches'. *Journal of Biological Education* **14**, 99–106
3.3   de Beer, Sir G. (1966) '200 Years of Malthus'. *New Scientist*, 17 February, 423–6
3.4   de Beer, Sir G. ed. (1958) *Charles Darwin and Alfred Russel Wallace: Evolution by Natural Selection*. Cambridge University Press
3.5   Edwards, K. J. R. (1977) *Evolution in Modern Biology* (Studies in Biology No. 87). Edward Arnold
3.6   de Beer, Sir G. (1956) Preface to the reprint of the sixth edition of *The Origin of Species* by Charles Darwin. Oxford University Press
3.7   Berry, R. J. (1982) *Neo-Darwinism* (Studies in Biology No. 144). Edward Arnold

# 4

# Variation: its origin and transmission

One of the principal defects of Darwin's argument was that he believed in a mechanism of inheritance which achieved the opposite of what his theory demanded. Thus, small variations among living organisms that were selected positively, far from being preserved, were destined to be progressively diluted and eventually lost as a result of blending. To overcome this difficulty, Darwin was driven to postulate the origin of fresh varieties at each generation – a view that conflicted with his field observations. Moreover, the idea was theoretically untenable for it implied that the bulk of variation available for selection at any moment must be of extremely recent origin. For it to survive, it would need to be selected almost instantly before becoming diluted by blending. In the face of such difficulties, doubts were beginning to be expressed by the end of the nineteenth century regarding the universal application of Darwinism as expressed in *The Origin of Species*.

Meanwhile, in 1866, a Silesian monk, Gregor Mendel (1822–84) (Fig. 4.1) working at Brünn in Moravia (now Brno, Czechoslovakia), had published his famous paper on inheritance in the garden pea, *Pisum sativum*. As we saw in the previous chapter (p. 50), interest in the mechanism of heredity at that time was predominantly biometric, being concerned with such aspects as the transmission of traits in families rather than the outcomes of specific experimental crosses. In the circumstances, it is not surprising that Mendel's work remained unnoticed for another 34 years until it was rediscovered, almost simultaneously, in 1900 by three botanists – H. de Vries in Holland, E. von Tschermak in Austria, and C. Correns in Germany. Its appeal to each of them was that it helped to explain the results they had obtained from the crossing of different plant varieties. Later, Mendel's findings were shown by William Bateson and R. C. Punnett in England, and L. Cuenot in France, to apply to animals as well.

## The novelty of Mendel's work

As a result of his experiments, Mendel showed that genetic factors (**genes**), which come together in a cross through sexual reproduction, do not blend but remain distinct and uncontaminated. In other words, inheritance is **particulate**, the genes behaving like particles. Moreover, apart from rare exceptions (**mutations**), the genes retain their identity in subsequent generations.

The work on peas established two important principles. The first of these is **segregation**: that genes derived from each parent segregate (separate) in the formation of the gametes of the offspring without affecting one another (Fig. 4.2). Thus, pairs of genes (**alleles**) which are separated in one generation are brought together in the next. However, an outcome of segregation is that the genes are able to reassort at random as in the $F_2$ generation (Fig. 4.2), thus increasing variation. Another important discovery of Mendel was the existence of **dominance**,

**Fig. 4.1**   Gregor Mendel in 1865, at the time of the publication of his paper on peas

whereby the effect of one gene (**dominant**) masked that of its allele (**recessive**). Thus, in the example, the gene combination (**genotype**) *Rr* produces a round pea (**phenotype**), not an intermediate between round and wrinkled. Since the pairing of gametes is at random, the probability of an experimental result equalling the theoretical value increases with the number of fertilisations. In the experiment shown in Fig. 4.2, the number of seeds produced was 7324 and we would have expected close agreement between the experimental ratio and that actually obtained.

Mendel extended the idea of segregation to situations involving more than one pair of alleles and showed that in the formation of gametes, the pairs assorted independently of one another. The principle is illustrated in Fig. 4.3. Compared with the previous example, it will be noted that in the $F_2$ generation the number of seeds is smaller (556) and the range of phenotypes greater, so we would not expect as close an agreement between observation and expectation as before. None the less, on statistical grounds, there is no reason to believe that the ratio obtained by Mendel diverged significantly from $9:3:3:1$.

We now know that independent assortment does not occur invariably, and where two or more genes are situated on the same chromosome they are **linked**

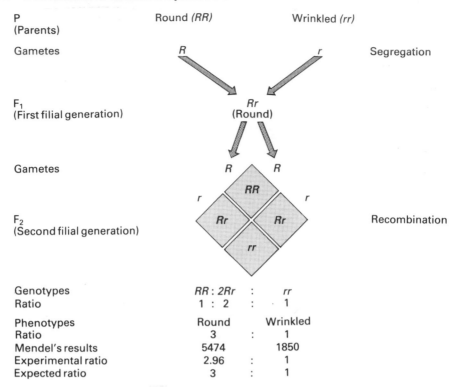

P
(Parents)          Round *(RR)*          Wrinkled *(rr)*

Gametes           *R*                    *r*          Segregation

F₁
(First filial generation)        *Rr*
                                 (Round)

Gametes                    *R*    *R*

                      *r*  ╱ RR ╲  *r*

F₂                   ╱ Rr ╳ Rr ╲        Recombination
(Second filial generation)
                          *rr*

| Genotypes | $RR$ : $2Rr$ | : | $rr$ |
|---|---|---|---|
| Ratio | 1 : 2 | : | 1 |
| Phenotypes | Round | Wrinkled | |
| Ratio | 3 | : | 1 |
| Mendel's results | 5474 | 1850 | |
| Experimental ratio | 2.96 | : | 1 |
| Expected ratio | 3 | : | 1 |

**Fig. 4.2**  The principles of segregation and recombination illustrated by Mendel with the characteristics 'round' and 'wrinkled' in peas

and therefore assort together. **Crossing-over**, another important genetic phenomenon, represents a breakdown of linkage. For further discussion of basic genetic principles reference should be made to the appropriate literature [4.1, 4.2]. In retrospect, we can see that luck was on Mendel's side in that he happened to select characters in peas that were not linked. In some of his later breeding experiments, for instance those of hawkweeds (*Hieracium*), he was less fortunate and encountered what he called 'peculiar behaviour'. We now know that this was due to **apomixis**, the formation of seeds without the fusion of gametes. None the less, his insight into the behaviour of genes and, in particular, their permanence laid the foundation of our modern understanding of **discontinuous variation**.

### The problem of continuous variation

Some of the variation we see around us is of the discontinuous kind; that is to say it is separable into distinct classes such as male and female, and normal colour and albino. However, many variants appear not to behave in this way but to form graded series, typical among them being human height and hair colour. As we saw earlier, at the time of Darwin a biometrical approach to heredity had tended to focus interest on this second kind of variation, usually known as **continuous variation**. In accordance with the views of the time, this agreed well with the expected outcome of a mechanism of blending inheritance. On the contrary,

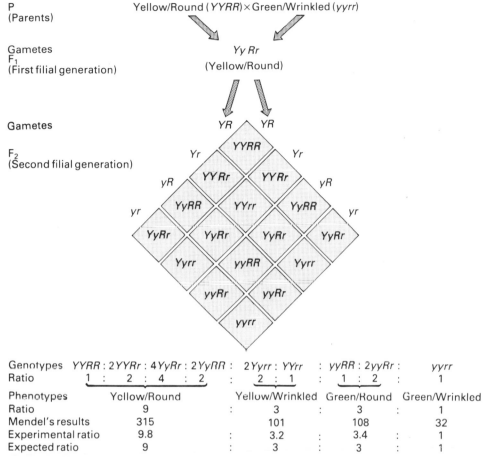

P
(Parents)
Yellow/Round (*YYRR*) × Green/Wrinkled (*yyrr*)

Gametes
F₁
(First filial generation)
*Yy Rr*
(Yellow/Round)

Gametes
F₂
(Second filial generation)

| Genotypes | *YYRR* : | 2*YYRr* : | 4*YyRr* : | 2*YyRR* : | | 2*Yyrr* : | *YYrr* | : | *yyRR* : | 2*yyRr* : | | *yyrr* |
|---|---|---|---|---|---|---|---|---|---|---|---|---|
| Ratio | 1 : | 2 : | 4 : | 2 : | | 2 : | 1 | : | 1 : | 2 : | | 1 |
| Phenotypes | Yellow/Round | | | | | Yellow/Wrinkled | | Green/Round | | | Green/Wrinkled | |
| Ratio | 9 | | | | : | 3 | : | 3 | | : | 1 | |
| Mendel's results | 315 | | | | | 101 | | 108 | | | 32 | |
| Experimental ratio | 9.8 | | | | : | 3.2 | : | 3.4 | : | | 1 | |
| Expected ratio | 9 | | | | : | 3 | : | 3 | : | | 1 | |

**Fig. 4.3** Independent assortment and inheritance of discontinuous variation in peas, with some of Mendel's results

Mendel's theory of particulate inheritance appeared to be directly opposed to any form of phenotypic continuity.

We now know the reason for this misunderstanding. It had not been appreciated that many characters such as the coat colour of mammals are controlled by several pairs of alleles, which are unlinked and additive in their effects. Moreover, the situation need not be complex in order to produce a graded series; only two pairs of alleles are sufficient.

Suppose we have a species of animal such as a mammal with a fur colour range of black through grey to white. Colour is controlled by two pairs of unlinked alleles, the dominants $B^1$ and $B^2$ promoting the production of the black pigment melanin and the recessives $b^1$ and $b^2$ inhibiting it. Each $B$ gene can therefore be regarded as contributing one unit of pigmentation; the maximum possible is four resulting from the genotype $B^1B^1B^2B^2$. Using the previous format, we can summarise the results of the F₁ and F₂ generations (Fig. 4.4). Presented in graphical form, the data conform to a typical bell-shaped curve of normal distribution (Fig. 4.5) comparable with that of other continuous variables such as human height or the weight of broad bean seeds.

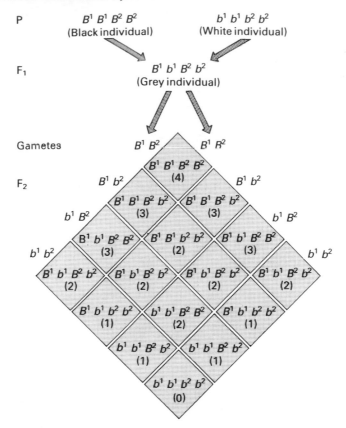

**Fig. 4.4** Variation in pigmentation resulting from the effects of two pairs of unlinked alleles. Numbers in brackets denote pigment units (theoretical example)

## What is a gene?

So far, we have assumed the existence of genes without attempting to define them. Now we must take our ideas a step further. A gene can be defined in a number of ways. Perhaps the simplest is to regard it as a unit of inheritance responsible for the control of one or more specific characteristics in an organism. Alternatively, we can think of it as the smallest unit capable of recombination or of undergoing spontaneous change (**mutation**). Again, we can think of it in a more functional sense as being responsible for initiating certain chemical processes leading to differentiation within a cell.

Following the rediscovery of Mendel's work in 1900, it was predicted two years later by W. S. Sutton, although not finally established for about another 20 years, that the chromosomes were the carriers of the genes. For the geneticists of the early twentieth century, the analogy was with rows of beads on a string. But as we saw in Chapter 1 (p. 4), following the contributions of James Watson and Francis Crick, and their successors, we now have a much clearer idea of the biochemistry of the chromosomes and the action of genes as parts of them. Thus, cells may contain two principal kinds of nucleic acid – deoxyribonucleic acid (DNA) and ribonucleic acid (RNA) – which consist of chains of units (**nucleo-**

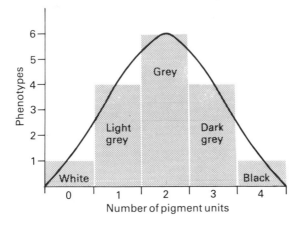

| Pigment units | Phenotype | Number |
|:---:|:---:|:---:|
| 4 | Black | 1 |
| 3 | Dark grey | 4 |
| 2 | Grey | 6 |
| 1 | Light grey | 4 |
| 0 | White | 1 |

**Fig. 4.5**  Inheritance of continuous variation. Graphical presentation of the data in Fig. 4.4

tides), each composed of a pentose sugar, organic base, and phosphoric acid (Fig. 1.2, p. 5).

The nucleotides of DNA are referred to by the names of their four bases which are adenine (A), thymine (T), cytosine (C), and guanine (G). It has now been established that the genetic code consists of a series of three-letter words or **triplets**. Now, four different symbols can be combined to form triplets in $4^3$ = sixty-four different ways, so the genetic alphabet permits the formation of sixty-four possible words – ATC, CGA, TAG, and so forth. A single gene may vary greatly in size from a dozen or so words to several thousands arranged in a linear sequence along the DNA chain. The chromosome complement of man ($2N$) is forty-six, and it has been estimated that the total genetic information carried by a single haploid gamete must amount to around 5000 million words. In a functional sense, we can assume the modern hypothesis to be that a single gene controls the production of a particular protein or part of it (the polypeptide). However, the situation is complicated by the finding that the portion of DNA concerned with the production of an individual protein may not be quite the same as that involved in a mutation. The implications of this need not concern us here.

## Mutation as the origin of variation

One of the essential requirements of particulate inheritance as a means of maintaining and transmitting variation is that the system should be stable and not subject to frequent alterations in the hereditary material. We would therefore predict that the molecules of DNA and RNA would reflect this property and that mutation would be a rare occurrence. We now know that both these predictions are justified.

## Gene mutation and variation

Experiments with a wide variety of plants and animals have shown that the average rate of **gene mutation** is of the order of one in a million individuals. Some genes mutate more frequently, such as that causing haemophilia in man where the rate is about one in 80 000. Again, in a particular species, the mutation rates of different genes vary considerably. Thus, in maize, *Zea mays*, genes controlling structure and pigmentation have been shown to mutate at frequencies ranging from 1.2 to 106 per million gametes. Although the mutation rate at a particular locus is low, it must be remembered that an individual gamete contains many genes (perhaps 10 000) so that the overall accumulation of new variants is, in fact, quite high. King [4.3] has suggested that as many as one gamete in ten may carry a mutant gene of some sort.

The development during recent years of much more sensitive methods for detecting gene mutation at the biochemical level, such as the various forms of electrophoresis, has established that the rates of change at some sites may be a good deal higher than had previously been believed possible. These are confined principally to microorganisms, but they have also been identified in higher animals such as vertebrates. Thus, in cattle, a gene controlling the blood group $\beta$-system mutates with a frequency of around 1 in 500. Again, in the brook trout, *Salmo fario*, a gene influencing the enzyme lactate dehydrogenase has a mutation rate of 1 in 50. Evidently, at molecular level, there are a large number of unstable genes that mutate repeatedly resulting in the accretion of a wide range of biochemical variation. To what extent such variants influence the survival of the organisms concerned is still largely unknown. Many of them may be selectively neutral or nearly so, and until we know more about their phenotypic effects it is difficult to judge their significance in the broader context of evolution.

At any site on a chromosome, a mutation may occur several times. Thus, a dominant gene $B$ controlling hair pigmentation might mutate to $B^1$ and $B^2$, altering appreciably the depth of colour of the resulting phenotype. Similarly its recessive allele $b$ could give $b^1$ and $b^2$. Gene series of this kind built up at a particular point on homologous chromosomes are known as **multiple alleles**. In the diploid individual, the gene pairs can occur in any combination such as $B^1b$, $Bb^2$, and so forth, the rule being that only one member from each chromosome can be included. The outcome in a population will be a graded range of pigmentation comparable to that discussed earlier (p. 59 and Fig. 4.4). However, the previous example depended upon the action of pairs of genes carried on *different* chromosomes, which therefore assorted independently; this is a mode of inheritance known as **polygenic**.

Bearing in mind the importance of a low mutation rate, it is not surprising to find that in some (possibly all) organisms this is itself under genetic control. Any gene tending to increase the mutation rate might be expected to be disadvantageous. Selection will therefore tend to act against it and it will become recessive to the normal form which will be dominant. Such a gene has been shown to occur in the fruitfly, *Drosophila melanogaster*, and to be carried on the second chromosome. Its effect in the homozygous state is to increase the mutation rate on the other chromosomes by a factor of 10, and in the heterozygous condition by a factor of between 2 and 7.

## Chromosome mutation and variation

Since the chromosomes are the carriers of genes, it follows that a multiplication of the number present in the nucleus of each cell (**polyploidy**) or alteration in the structure of the chromosomes themselves (**fragmentation**) may have profound effects on the organisms concerned and result in new varieties. One of the commonest abnormalities is an increase in the chromosome number (polyploidy). In the sexual reproduction of diploid organisms, the parental cells contain the double number of chromosomes $(2N)$. Following meiotic division in gametogenesis, gametes are produced with the haploid chromosome number $(N)$. At subsequent fertilisation and the production of the next generation, the diploid number is restored. The sequence of events for two chromosomes $A$ and $B$ can be represented as:

$$\text{Parents } AABB \times AABB \quad (2N)$$
$$\text{Gametes } AB \times AB \quad (N)$$
$$\text{Fertilisation } AB \times AB \rightarrow AABB \quad (2N)$$

One way in which this sequence of events can become upset is by the failure at meiosis of a pair of chromosomes to separate, resulting in gametes with different chromosome complements. This is known as **non-disjunction** and can be represented as follows using the same notation as before:

$$\text{Parents } AABB \times AABB \quad (2N)$$
$$\text{Gametes } AAB \times B$$
$$\text{Fertilisation (by normal gamete) } AAB \times AB \rightarrow AAABB \text{ (trisomic)}$$
$$B \times AB \rightarrow ABB \text{ (not viable)}$$

In this instance, only gene $A$ is increased from $2N$ to $3N$; the remainder are unchanged. Such a condition is referred to as **trisomic**. Had the numbers of all the genes increased to $3N$, the result would have been a **triploid**. One of the best-known examples of trisomy following non-disjunction is Down syndrome (mongolism) in man. Normal human development requires two sex chromosomes and twenty-two pairs of autosomes (i.e. $2N = 46$). But in Western Europe, one birth in 2000 to mothers aged 29 and over suffers from non-disjunction of chromosome number 21, resulting in trisomy. The phenotypic effects constitute Downs syndrome and include extreme mental retardation, various physiological difficulties such as heart defects, and a low expectation of life.

Another form of polyploidy can result from the failure of chromosomes to separate at mitosis of the zygote, giving rise to doubling, trebling, and so forth. The process can be illustrated as:

$$\text{Parents } AABB \times AABB \quad (2N)$$
$$\text{Gametes } AB \times AB \quad (N)$$

$$\text{Fertilisation } AB \times AB \rightarrow AABB \xrightarrow[\text{mitosis}]{\text{doubled at}} AAAA\,BBBB \quad (4N)$$

Polyploids derived from a simple process of doubling are known as **autopolyploids**; if they contain $4N$ chromosomes they are **autotetraploids**. Usually, such individuals are fertile with one another, but since they have a different chromosome number from normal members of the stock they are unable to form compatible gametes and are therefore infertile with them.

It must be remembered that autopolyploids add no new alleles to the existing gene complement; they only multiply those that exist already. As might be expected, their predominant effect on the phenotype is to enhance characteristics already present rather than adding new ones. Polysomics and polyploids are common among plants where they have been used commercially to produce many important domestic strains (**cultivars**). Among wild species, the thornapple, *Datura stramonium*, has produced a range of chromosome variants including tetraploids, trisomic diploids, and pentasomic tetraploids (Fig. 4.6). The adaptive significance of the different forms is uncertain but they can be distinguished from one another by the shape and size of the seed capsules (Fig. 4.6). In general, the greater the value of $N$ the larger the capsule, but this is modified by a single gene ($G$) which, in varying doses, modifies capsule shape and the size of the spines.

In a number of species, the values of $N$ are simple multiples of one another. Thus, the narrow-leaved bird'sfoot trefoil, *Lotus tenuis*, with slender growth and small flowers, has a diploid number of 12, whereas that of the much more robust common bird'sfoot trefoil, *L. corniculatus*, is 24. Sometimes, polyploidy appears to have a more clear-cut adaptive significance as in the marsh bedstraw, *Galium palustre*. While the normal diploid ($2N = 24$) occurs in damp places which are subject to periodic drying, the tetraploid ($4N = 48$) is found in wetter localities such as the edges of streams that remain moist throughout the year. The octoploid ($8N = 96$) is also known and inhabits sites similar to those of the tetraploid. The situation in *Galium* appears to be typical of a general tendency among plants – the greater the chromosome number, the better the ability to colonise damp habitats. Indeed, the different chromosome races are sometimes so similar in appearance that the best way to separate them is by their differing ecological requirements.

As was illustrated in the thornapple, *Datura stramonium* (Fig. 4.6), polyploidy and polysomy frequently result in chromosome numbers of a species which are not simple multiples of one another. These are known as **aneuploids**. A typical

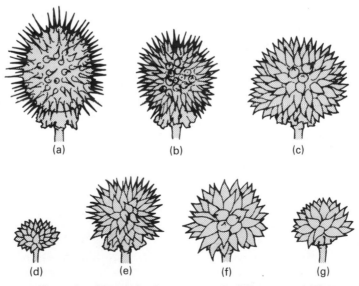

**Fig. 4.6** Polyploidy and polysomy in the thornapple, *Datura stramonium*, and their influence on the size and shape of the seed capsule: (a) $2N$, (b) $4N$, (c) $2N + G$, (d) $2N + 2G$, (e) $4N + G$, (f) $4N + 2G$, (g) $4N + 3G$ (from King 1962)

example is in the cuckoo flower, *Cardamine pratensis*, where a diversity of chromosome numbers ranging from 30 to 76 exists. As in *Galium*, the higher the value of $N$, the greater appears to be the ability to colonise wet conditions. It has been estimated that as many as a third of all flowering plants may be polyploids or polysomics of some kind, which suggests that the adaptive advantages associated with increased chromosome numbers, such as those outlined above, may be more widespread than we realise.

In agriculture and horticulture, polyploids are equally widespread and cultivars of great commercial value range from cereals and sugar-beet to tomatoes and roses. The use of substances that increase the mutation rate (**mutagens**) has proved valuable in artificially speeding up the production of new variants. One of the most important of these is **colchicine**, an extract of the meadow saffron, *Colchicum autumnale*. Its effect is to inhibit the formation of the spindle at mitosis, with the result that the chromatid pairs fail to separate at anaphase and the chromosome number is therefore doubled.

There is a second way in which polyploids can be formed. Provided two species are fairly closely related, it is sometimes possible to cross them and produce hybrids. This is particularly true in plants. Hybrids are usually infertile because they possess only one set of chromosomes derived from each parent so, at meiosis, there are no homologous chromosomes with which pairing can take place. However, should doubling occur in the chromosome number of the hybrid, the formation of gametes then becomes possible. Suppose we have two chromosomes $A$ and $B$. The sequence of events would be:

$$\text{Parents } AA \times BB$$
$$\text{Gametes } A \times B$$

$$\text{Fertilisation (hybrid) } AB \xrightarrow[\text{mitosis}]{\text{doubled at}} AABB$$

$$
\begin{array}{ccc}
\text{Gametes} & - & AB \\
& \text{(sterile)} & \text{(fertile)}
\end{array}
$$

Polyploids resulting from chromosome multiplication of a hybrid are known as **allopolyploids** – in this instance an **allotetraploid**. Like autopolyploids, these have been widely used in commerce and their rate of production can be greatly increased by mutagenic agents such as colchicine and ionising radiations. One of the classic examples was obtained by G. D. Karpechenko as long ago as 1928 when he crossed a radish, *Raphanus sativa* ($2N = 18$), with a cabbage, *Brassica oleracea* ($2N = 18$). The fact that both happen to have the same chromosome number is coincidental; it does not signify that the two sets of chromosomes are homologous. The resulting semisterile hybrid possessed thirty-six chromosomes and produced a few seeds from which a further generation was raised, also with thirty-six chromosomes, which was fertile and vigorous. Named *Raphanobrassica*, this combined many of the characteristics possessed by the original parents.

The extent to which allopolyploidy has contributed to evolutionary changes in wild populations is uncertain but a well-documented case deserves repetition. This concerns the cord grasses (*Spartina*), a group of plants which habitually colonise the mud flats of estuaries. The indigenous British species is the lesser cord grass, *S. maritima*, whose distribution is widespread but localised. In 1878, there was a report of a hybrid in Southampton Water between *S. maritima* ($2N = 60$) and the American species *S. alterniflora* ($2N = 62$). The circumstances in

**Fig. 4.7**  The cord grass, *Spartina townsendii*, colonising the mud of a tidal estuary

which hybridisation occurred are unknown but *S. alterniflora* or the hybrid could well have been introduced by chance in the cargo of ships. The sterile hybrid was found to contain 62 chromosomes and subsequent doubling produced numbers of 124 and sometimes 120, with restored fertility. Back-crosses from these to the parent *S. maritima* (62 + 30) appear to have given rise to plants with a chromosome complement of 92. From these, a further back-cross (46 + 30) accounts for plants with a 2*N* value of 76. This is a new species, *S. townsendii* (Fig. 4.7), fertile with itself but sterile with both its original parents. Compared with *S. maritima* and *S. alterniflora*, it is better adapted to estuarine conditions, having a more vigorous growth and a greater ability to withstand fluctuating levels of salinity. Today, many of our estuaries and harbours are fringed with vast growths of *S. townsendii*, which, if unchecked, can reach pest proportions.

## Chromosome fragmentation and variation

Just as variations in the number of chromosomes in each cell nucleus can influence the constitution of an organism, so can the number of genes carried on each chromosome and their locations. In order to establish such relationships, it is necessary to be able to identify the positions of genes in relation to some structural characteristics of chromatin. Such an approach has been greatly aided by the discovery that flies (Diptera) such as the fruitfly, *Drosophila melanogaster*, possess giant chromosomes in their salivary glands with a characteristic banded appearance (Fig. 4.8). To the experienced observer, changes in the sequence or number of the bands are readily detectable.

Fragmentation in chromosomes can have four possible outcomes: the removal

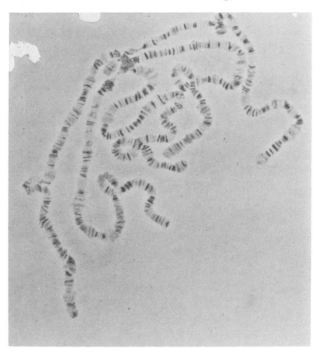

**Fig. 4.8** A giant chromosome of the fruitfly, *Drosophila melanogaster*. The sequence of bands can be used to identify the positions of genes

of a block of genes (**deletion**), a change in the gene sequence (**inversion**), or an alteration in the number of genes per chromosome (**translocation** and **duplication**).

*Deletion*

The principle is illustrated in Fig. 4.9(a) and involves the loss of a portion of one chromosome of a homologous pair, together with the genes it carries. Such occurrences are well known in laboratory stocks of *Drosophila*, following the painstaking mapping by C. B. Bridges and L. Slizynski of the positions of the 5149 bands on the four giant salivary gland chromosomes (Fig. 4.8). Thus, a deletion occurring at one site removes the segment of the chromosome carrying the gene for *notch*. Its absence can be detected by the characteristic shape of the tip of the wing. The significance of deletions in wild populations of plants and animals is still largely unknown. As with most mutations producing observable effects, these usually seem to have negative survival value. Thus in man, a form of leukaemia is associated with a cytological condition in which a member of one chromosome pair is shorter than the other, evidently due to the deletion of the missing portion.

*Inversion*

Sometimes a portion of a chromosome may become detached and later re-connected, but the other way round. The sequence of genes in that section therefore becomes reversed (Fig. 4.9(b)). The most elegant study of the evolutionary significance of inversions was carried out by T. Dobzhansky and his

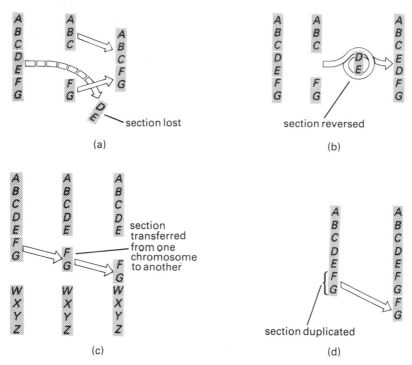

**Fig. 4.9** Four kinds of chromosome fragmentation: (a) deletion, (b) inversion, (c) translocation, (d) duplication. Each results in a change in the number of genes (indicated by letters) or their order, or both

colleagues on the fruitfly, *Drosophila pseudoobscura,* and other species in different areas of North America. Cytological evidence for chromosome abnormality was derived from the banding on the giant chromosomes. Evidently, the frequency of these inversions varies from one locality to another and at different times of the year. This suggests balanced ecological advantages and disadvantages of the various phenotypes which are subject to the influence of natural selection. The inverted gene sequences were named after the locality where each was discovered. Thus, the frequency of *standard* was high in March, at a minimum in June and thereafter rose steadily to a maximum in October. *Chiricahua,* on the contrary, behaved in exactly the opposite way with minimum frequencies in March and October, and a maximum in June. *Standard* chromosomes, with their particular gene sequence, were at an advantage during cool conditions whereas *Chiricahua* did best during the hottest period. This hypothesis was tested in the laboratory using a mixed population of the heterozygotes *Standard* and *Arrowhead.* It was found that at 16 °C all the *Standard* were eliminated, whereas at 25 °C an equilibrium was attained between the two.

As Ford has pointed out [4.1], where the heterozygotes showed superior survival value to the normal homozygotes, the flies all came from a particular locality. But when a cross was made between individuals caught 1000 km apart, the advantages of the heterozygotes were no longer apparent and the homozygotes were superior. Thus, the heterozygous condition possessed no inherent advantage of its own but only predominated when selected positively in particular situations.

As explained earlier, the existence of the various inversions is only detectable cytologically, so it is still not clear by what means they exert their phenotypic effects. Unlike the Mendelian systems considered previously (p. 56), the genetic mechanism here is not a single pair of alleles but a block of genes closely linked and operating as a kind of switch mechanism. Such an arrangement is known as a **supergene**; and since the genes involved are so close together, crossing-over in that portion of the chromosome is almost impossible.

*Translocation*

This involves the detachment of a portion of one chromosome and its attachment to another (Fig. 4.9(c)). Unlike the two processes considered earlier, this produces changes in two chromosomes, one an addition and the other a reduction. Such cytological upsets are likely to have significant effects on meiosis in gametogenesis where certain chromosome sections will be missing and others occurring in duplicate. Little is known about the occurrence of translocations in animals outside the laboratory, although it has been established that a form of mongolism in man somewhat resembling Downs syndrome (p. 63) can be the result of a translocation. Like all mutations, their rate of occurrence can be greatly increased by ionising radiations and these have been used by C. Patterson and others to study translocations in the fruitfly, *Drosophila*. One finding has been that a high level of lethality may occur in the homozygous condition. Thus, out of 120 translocations between chromosomes 2 and 3, only 19 were viable as homozygotes.

Translocations are widely known among plants, one of the most extensively studied being in the evening primrose, *Oenothera lamarckiana*, which possesses a translocation complex involving at least two linkage groups (supergenes) where crossing-over is suppressed. As in *Drosophila*, there is a high level of lethality among homozygotes leading to increased heterozygosity. Now it is well known among plant breeders that cultivars heterozygous at a large number of gene sites (**loci**) tend to exhibit a condition of **hybrid vigour** or **heterosis**. This improves such features as growth rate, reproductive capacity, and general viability when compared with those of homozygotes. It has been suggested that this greater biochemical versatility may result from a wider diversity of genes and hence the likelihood of an increased range of enzyme systems being produced. As King [4.3] has suggested, plant species such as *Oenothera lamarckiana* have evolved genetic systems of heterozygosity where translocation has been used to advantage. The result has been a more efficient metabolism and therefore greater adaptability. It seems likely that *O. lamarckiana* must have evolved from its close relative, *O. hookeri*, and it is, perhaps, significant that in this species translocation is absent.

*Duplication*

As a result of an abnormality of cell division, one or more segments of a chromosome may become duplicated as is illustrated in Fig. 4.9(d). The process is sometimes referred to by the alternative name of **repetition**, the reduplicated chromosome segments being called **repeats**. Little is known of its natural occurrence but it has been studied in some detail under laboratory conditions in the fruitfly, *Drosophila melanogaster*. Thus, the gene *Bar* influences the development of the eye by controlling the rate of cell division. Sutton showed that the segment of the chromosome containing the *Bar* gene may be represented once, twice, or three times (Fig. 4.10), giving the range of genotypes illustrated in Fig. 4.11. It will be seen that as the number of doses increases the number of facets in

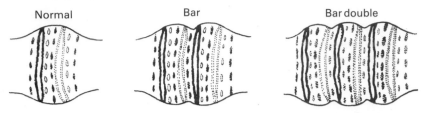

**Fig. 4.10** Repetition at the *Bar* eye locus in *Drosophila melanogaster* (after Sutton 1936)

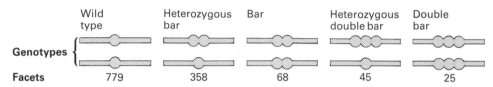

**Fig. 4.11** Possible *Bar* genotypes and the average number of facets in the compound eyes of *Drosophila melanogaster* (from King 1962)

the eye decreases, the highest and lowest values differing by a factor of about 30. A further point of interest is that heterozygous double *Bar* individuals have fewer facets than homozygous *Bar*, even though they both carry the same number of doses of the gene. Evidently, not only is the number of repeated chromosome segments important, but also their position. This has been called the **position effect**, and is important in highlighting the fact that the expression of a gene may depend not only on the external environment in which an organism lives (e.g. temperature and nutrition), but also on the internal environment provided by the other genes in the nucleus of the cell.

The number of facets in the compound eye of an insect could have considerable survival value, but, as indicated earlier, little is known of the evolutionary significance of this and other instances of duplication. Yet the process has great potential importance for, apart from polyploidy (p. 63), it is the only mechanism known at present whereby the *number* of genes possessed by a species can be increased.

As a final comment on chromosome fragmentation in general, it is worth noting that its occurrence is far more frequent in plants than in animals, where it has been identified mainly among laboratory stocks of species such as *Drosophila*. The reason for this disparity may well be that any change involving whole blocks of genes is likely to cause a considerable upset in the internal environment of the genotype. The structural and physiological make-up of plants is simpler than that of animals and may therefore be able to withstand such occurrences more effectively and even to profit from them. By contrast, the complex homeostatic systems of animals are more sensitive to genetic upset and therefore respond less favourably.

## Non-nuclear inheritance

The ratio of the volume of cytoplasm in a cell to that of the nucleus can be large; this is particularly true of female gametes where it frequently attains a level of

500:1 or more. It is therefore not surprising to find that in some circumstances influences within the cytoplasm can exert hereditary effects independent of those carried by the chromosomes. As might be expected, the different kinds of non-nuclear inheritance constitute a diverse assemblage which can be classified roughly as follows.

## Maternal inheritance

Many instances have been described among both plants and animals where the influence of the cytoplasm modifies or overrides that of the nucleus. A typical example is the inheritance of shell coiling in the wandering snail, *Limnaea pereger*, a common inhabitant of fresh water. This is controlled by a pair of alleles ($R$ and $r$) right-hand (dextral) coiling being dominant to left (sinistral). However, the phenotypic outcome is, in fact, determined through the maternal cytoplasm of the egg. Thus, a heterozygous dextral mother ($Rr$) crossed with a $Rr$ or $rr$ father will produce a half $rr$ offspring. These are genotypically sinistral but since their mother was dextral they are dextral too. On the other hand, $rr$ offspring from a sinistral female will be left-handed in accordance with their genotype. Thus, the genotype of the mother's cytoplasm determines the phenotype of the offspring.

## Chloroplast inheritance

This is widely known among plants where the egg is responsible for transmitting degrees of greenness to which the pollen makes no contribution. In some species, plants with green leaves produce only green progeny, yellow give yellow and variegated (mixed colours) have only variegated offspring. A typical example of this kind of cytoplasmic inheritance is the transmission of leaf colour in *Primula sinensis*, which is entirely non-nuclear. Thus:

| ♂ | × ♀ | Offspring |
|---|---|---|
| Yellow | × Green | ⟶ Green |
| Green | × Yellow | ⟶ Yellow |

Many similar examples are known, particularly among cereals. How the cytoplasm exerts its hereditary effects is still uncertain; indeed it may do so in several different ways. Chloroplasts are usually, but by no means always, transmitted from one generation to the next through the maternal line, suggesting that they are not necessarily required for hereditary characteristics to be handed on via the cytoplasm. On the other hand, they are able to undergo fission, to mutate, and reproduce the mutant forms. They are also known to contain DNA and RNA, so are equipped to transmit genetic information.

## Mitochondrial inheritance

In some ways, mitochondria resemble chloroplasts in that both have the capacity for transmitting hereditary information. This is well illustrated in baker's yeast, *Saccharomyces cerevisiae*, where a dwarf form (*petite*) has long been known. This occurs in many populations at a frequency of around 1 per cent and can multiply by budding (vegetative reproduction) producing further *petite* populations. The dwarf form has been shown to lack various respiratory enzymes due to the loss of mitochondria. B. Ephrussi obtained a French strain of yeast whose characteristics resembled *petite* but were controlled by a recessive gene which inactivated the mitochondria. On crossing with the original *petite* stock containing a wild-type gene, the inactivated mitochondria recovered their activity and normal growth occurred. Thus, the synthesis of the necessary respiratory enzymes depended upon an interaction between the genotype and cytoplasmic inclusions. A similar situation exists in the mould *Neurospora*, where a mutant form known as *poky*, characterised by its slow growth, is similarly under the interactive control of mitochondria and the nucleus.

**Cytoplasmic hereditary particles**

Apart from chloroplasts and mitochondria, a range of other bodies have been identified in the cytoplasm of cells and shown to have hereditary properties independent of those of the nucleus. Some of the most remarkable are the **kappa particles** discovered by T. M. Sonneborn in the ciliate, *Paramecium aurelia*. These contain DNA and protein, and electronmicrographs have shown them to have a complex internal structure. They are distributed freely in the endoplasm of the animal and during their early development are capable of division and growth, eventually reaching a size of about 0.8 × 2.3 μm. Individuals that do not possess kappa particles are sensitive to a toxic substance (paramecin) liberated by those that carry them, the effect being to cause abnormal morphology and behaviour, followed by death in a few hours. Animals possessing kappa particles (**killers**) are somehow protected by the inclusions from the effects of the toxins they produce.

The capacity for infection and multiplication of kappa is controlled by a dominant nuclear gene $K$, which cannot initiate production but promotes it once infection has occurred from the outside. In the recessive state ($kk$), the individuals may contain particles originally but they eventually disappear and the animals become sensitives.

The number of kappa particles per individual varies from 100 to 3000 and, if an animal is to remain a killer, it is vital that binary fission of the organism and multiplication of kappa should keep in step. In experimental conditions, the rate of reproduction of *Paramecium* can be made to exceed that of the particles, with the result that their numbers gradually decline and a kappa-free population of sensitives is left. The existence of kappa particles in *Paramecium* thus provides a beautiful example of **symbiosis**, in which a unicellular animal and an infective symbiont have become genetically and ecologically co-adapted in an intricate and mutually beneficial manner.

# Environmental variation

The morphological appearance of a plant or animal depends upon the interaction between the organism's genes and the environment. As we have seen, the gene environment consists partly of the genotype as a whole (**internal environment**) and partly of the physical conditions in which the species exists (**external environment**). It has been estimated that of the total variation exhibited by a species, as much as half may be genetic (and therefore inherited) and the remainder environmental (and therefore not inherited).

Some of the most obvious environmental influences occur during the period of development. Thus, in the Siamese cat the gene controlling the typical pale-brown fur colour depends for its expression on the body being maintained at a constant temperature. At birth, the kittens are uniformly coloured, having developed within the maternal uterus. But once exposed to the outside, the extremities cool below the critical temperature and the hair on the nose, ears, paws, and tip of the tail grows dark brown. If a paw of a newly born kitten is wrapped in an insulating material such as flannel and kept warm, the change of colour can be prevented. Similarly, if a patch of light-coloured hair is shaved and then cooled, dark fur grows; but it is gradually replaced by its former colour once the typical temperature conditions are restored. Again, among butterflies, adults of the small tortoiseshell, *Aglais urticae*, reared in warm conditions are typically light coloured, while those kept in the cold are much darker with many black scales.

Plants, too, exhibit similar environmental variations to an even greater degree. For instance, species growing on coastal sand dunes have to adjust to a considerable salinity in the soil and a shortage of water. One way in which this is achieved is by increasing the amount of water storage tissue in the leaves giving a fleshy appearance, as in some common weeds such as the scarlet pimpernel, *Anagallis arvensis*, where plants from dunes and gardens are easily distinguishable. One of the most powerful environmental influences on plants is that of light. Thus, a cabbage lettuce grown in full light has a rounded appearance with a compact central 'heart'. But grown in the shade, the leaves of the plant are elongated and practically no heart is formed. The effects of illumination can cause considerable difficulties in identifying wild plants if only vegetative characters are used. Thus, the field scabious, *Knautia arvensis*, is a plant of open grassland and maximum sunshine. In such conditions, its leaves are of the type shown in Fig. 4.12(a); but in the shade of a bush such as hawthorn, the leaves can assume the appearance of Fig. 4.12(b) and be almost unrecognisable as belonging to the same species.

Buffon, Lamarck, and even Darwin to some extent, thought that environmental variants of the kind described above could be incorporated in the genetic constitution of an organism and inherited. But with the coming of the germ theory of Weismann (p. 49), it was clear that the germ cells associated with sexual reproduction and the somatic cells of the rest of the body were quite distinct. In spite of many attempts to prove the contrary, there is still no evidence that somatic modifications resulting from the action of the environment during the lifetime of a plant or animal can become part of its hereditary make-up.

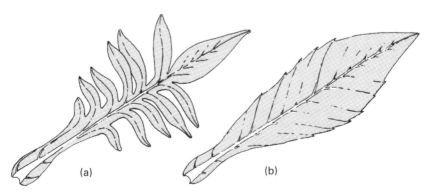

**Fig. 4.12** Environmental variation in the field scabious, *Knautia arvensis*: leaves of plants growing in (a) full sun, (b) the shade

## Variation in populations

Mendel's theory of particulate inheritance was concerned with the results of individual crosses. But all living organisms exist in **populations** – interbreeding groups consisting of the same species. So the question soon arose whether Mendelian principles were applicable to these as well. In 1908, writing in the *Proceedings of the Royal Society of Medicine*, G. U. Yule suggested that if brachydactyly (shortened middle fingers and toes) in man is dominant 'in the course of time one would expect, in the absence of counteracting factors, to get three brachydactylous persons to one normal'. This question posed a fundamental problem. Does the dominance or recessiveness of a gene determine the frequency of its occurrence in a population? The question was answered in 1908 by the English mathematician G. H. Hardy in one of the shortest ($1\frac{1}{2}$ pages) yet most far-reaching scientific papers ever written [4.4]. Incidentally, at the beginning of his paper, he makes a splendidly dismissive comment on biologists without mathematics – 'I should have expected the very simple point which I wish to make to have been familiar to biologists.' By a curious coincidence a German physician W. Weinberg pointed out much the same facts at almost the same time, so the principle they enunciated came to be called the Hardy–Weinberg law. It provides an essential part of the foundation of population genetics as we know it today.

Suppose two alleles $A$ and $a$ occur in a population and the frequence of occurrence of $A$ is $p$, and that of $a$ is $q$. Then $p + q = 1$ (i.e. 100 per cent of the population). Three genotypes are possible, $AA$, $Aa$, and $aa$. The outcomes of the possible gene combinations and their frequencies are summarised in Fig. 4.13.

Genotype $AA$ results from the union of two gametes both carrying $A$. Since the frequency of $A$ is $p$, the probability of its occurrence is $p \times p = p^2$. Genotype $aa$ similarly will have a frequency of $q \times q = q^2$. Genotype $Aa$ differs from the others in that $A$-bearing gametes may be contributed by either male or female, similarly the $a$-bearing gametes (see Fig. 4.13). There are therefore two possibilities:

$$\begin{aligned} (A \times a) &= p \times q = pq \\ (a \times A) &= q \times p = pq \end{aligned} = 2pq$$

or

The three genotype frequencies must add up to 1 (i.e. 100 per cent). Therefore, we can write,

$$p^2 + 2pq + q^2 = 1$$

This is the Hardy–Weinberg equation.

Returning to the example in Fig. 4.13, we can rewrite the equation as:

$$p^2_{(AA)} + 2pq_{(Aa)} + q^2_{(aa)} = 1$$

Thus, the proportion of $A$ alleles in the $F_2$ generation is,

$$\frac{p^2 + pq}{p^2 + 2pq + q^2} = p^2 + pq = p^2 + p(1 - p) = p$$

Similarly, the frequency of $a$ in the $F_2$ generation is $q$.

Returning to the proposition of Yule quoted earlier, there was, therefore, no justification for his assumption that the incidence of brachydactyly would move towards Mendelian proportions in subsequent generations. In the absence of other modifying factors, the frequency of the gene remains constant at each successive generation.

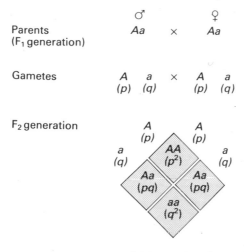

$$= p^2\,AA\,;\,2pq\,Aa\,;\,q^2\,aa = 1\,(\text{total of zygotes})$$

**Fig. 4.13** Derivation of the Hardy–Weinberg ratio. Genes $A$ and $a$ occur in a population with a frequency of $p$ and $q$ respectively

The Hardy–Weinberg principle depends upon simple probability and the chance of two variables combining at random in particular ways within a **closed system**. In population genetics such a system is known as a **gene pool**. To provide such a situation, certain requirements must be fulfilled:

(i) All genotypes must have an equal chance of mating (i.e. fertilisation must be at random).

(ii) The size of the population must be sufficient to exclude effects due to non-random mating.

(iii) The three genotypes must have equal chances of survival (i.e. there must be no selection of the resulting phenotypes). Moreover, there must be equal chances of reproduction (i.e. all must have the same fertility).

(iv) Mutation must be sufficiently rare to have no significant effect on the frequencies within the gene pool.

(v) There must be no immigration or emigration (i.e. the population must be reproductively isolated).

The chances of a Hardy–Weinberg equilibrium existing within a population over a long period are, of course, remote. But for short periods of a few generations, approximate stability of gene frequencies may prevail. The important point is that, should frequencies not conform to expectation, this is evidence of the action of one or more of the factors outlined above. The value of Hardy–Weinberg, therefore, is that it serves to indicate the existence of factors influencing gene frequency although it does not identify the particular agents involved.

An example will illustrate the approach in an actual context. The human blood condition known as sickle-cell anaemia is widespread in Africa, parts of India and in some areas of the Mediterranean. It is caused by a haemoglobin variant under the control of a pair of alleles which we can call $A$ and $S$. The homozygotes ($SS$) are severely affected and many of them die. The heterozygotes ($AS$) are also affected but less so, and most of them survive. The two can be distinguished by means of blood smears and also biochemically by electrophoresis. In some parts of Africa, the proportion of homozygous individuals is 4 per cent. What level of heterozygotes would we expect if we assume a Hardy–Weinberg equilibrium? We can write:

| Genotypes | $AA$ | $AS$ | $SS$ |
|-----------|------|------|------|
| Proportions | $p^2$ | $2pq$ | $q^2$ |

The incidence of homozygous individuals $(SS) = q^2 = 0.04$ (i.e. 4 per cent).

Now, $p + q = 1$ where $p =$ the frequency of $A$ and $q =$ the frequency of $S$. Thus:

$$q = \sqrt{(0.04)} = 0.2$$
$$p = 1 - 0.2 = 0.8$$
$$p^2 = (0.8)^2 = 0.64 \ (64 \text{ per cent})$$
$$2pq = 2 \times 0.8 \times 0.2 = 0.32 \ (32 \text{ per cent})$$

So in a population containing 4 per cent homozygotes, we would expect 32 per cent to be heterozygous for the sickle-cell gene. In fact, the level is nearer 40 per cent, suggesting that more individuals survive than would be expected to do so. The sickle-cell condition has now been subjected to detailed analysis by A. C. Allison and has revealed a fascinating instance of the balance of selective advantage against disadvantage. The findings discussed here will be considered again in their evolutionary context in the next chapter.

Finally, it must be stressed that the incidence of sickle-cell anaemia provides a particularly good example of the use of the Hardy–Weinberg principle as a diagnostic instrument, since the three classes of phenotypes are easily distinguishable. More often, homozygous dominants resemble heterozygotes so only two classes can be identified. However, the use of techniques such as electrophoresis of body fluids are gradually providing new approaches to such problems of genetic analysis.

## Summary

1 The discoveries of Mendel overcame one of the problems facing Darwin. In propounding a mechanism of particulate inheritance, he disposed of the need to postulate blending.

2 The processes of segregation and independent assortment of genes provide a means of maintaining and transmitting variation.

3 Both discontinuous and continuous variation are equally explicable in Mendelian terms.

4 Genes can be defined in a variety of ways. Discovery of the genetic code has greatly increased our knowledge of gene action.

5 Mutations are the source of new inherited variation; their occurrence is rare. Mutants can occur at the site of a particular gene or as a result of chromosome fragmentation.

6 Polyploidy and polysomy have been particularly important in the commercial production of new strains of plants and animals. Mutagenic agents such as colchicine and ionising radiations have also been used to produce new mutants.

7   Various kinds of non-nuclear inheritance are found, particularly in plants. Compared with the nuclear process, they have played a relatively unimportant part in evolution.

8   It has been estimated that roughly 50 per cent of all variation in plants and animals may be environmental. In this connection, it is important to distinguish between the internal and external environment.

9   Environmental modifications acquired during the lifetime of an organism, such as changes in shape and structure, are not inherited.

10   The Hardy–Weinberg principle provides a simple mathematical model for studying the maintenance and transmission of variations in populations. It provides a useful means of detecting the action of natural selection and other factors influencing gene frequencies.

11   In the absence of modifying factors, the frequencies of genes in a gene pool remain constant irrespective of dominance and recessiveness.

## Topics for discussion

1   Trace the course of a gene mutation from the time of its occurrence until it reaches a frequency of, say, 50 per cent in a population.

2   Two different processes are claimed to have contributed to the production of new species – the cumulative effect of numerous small mutations, and fewer macromutations with large phenotypic effects. On theoretical grounds, what are the arguments for and against each process?

3   It is possible for two dark-haired parents to have a red-haired child. What is the genetic mechanism involved?

4   What is the evidence that deoxyribonucleic acid (DNA) is an integral part of the gene?

5   In what circumstances could a mutant gene which was originally recessive become a dominant?

6   Explain in detail why the dominance of a gene does not ensure its spread through a population?

7   Construct a model to show how the continuous variation of a quantitative characteristic such as coat colour in mammals conforms to Mendelian principles.

8   Why is non-nuclear inheritance likely to have played only a limited role in evolution?

9   The Hardy–Weinberg formula provides a good example of the limitations of a simple mathematical model. What are its limitations?

10   If 'like tends to beget like', how do you correlate this with the fact that evolution has occurred?

## References

4.1   Ford, E. B. (1979) *Understanding Genetics*. Faber
4.2   Kemp, R. (1970) *Cell Division and Heredity* (Studies in Biology No. 21). Edward Arnold
4.3   King, R. C. (1962) *Genetics*. Oxford University Press
4.4   Peters, J. A. ed. (1959) *Classic Papers in Genetics*. Prentice-Hall, Englewood Cliffs, New Jersey

# 5

# Natural selection at the species level

From all his writing, it is clear that Darwin viewed the species as the natural point to begin the study of evolutionary change. His view has been amply upheld by subsequent experience. So it is to the species level and below that we must turn if we are to learn something of the process of natural selection in action.

One of the merits of Darwinism is that it accounts for differential survival and change in terms of a variable which, in theory at least, can be identified and quantified. This is **adaptation**. Thus those individuals best adapted to their ecological environment will, on average, tend to survive and breed at the expense of the less well endowed. Before pursuing this idea further, we need to be clear as to what we mean by the term **environment**.

## The nature of the environment

The environment in which a plant or animal lives represents the total of the ecological conditions that exist there. These are of two kinds. Physical (**abiotic**) factors are of a climatic kind such as light, temperature, and humidity. Living (**biotic**) factors are the outcome of interactions with other living things, such as food (e.g. predation and parasitism), space (e.g. population density), and competition, both within species (**intraspecific**) and between them (**interspecific**).

An essential feature of all environments is that they are dynamic systems in a constant state of **fluctuation**. Sometimes these fluctuations occur with a regular periodicity but more often they are erratic and unpredictable. Their origin may be **extrinsic** such as light, temperature and food supply, which are subject to seasonal change. But they may also be **intrinsic** and an outcome of the activities of the organisms themselves, such as changes in the illumination in a wood as a result of the growth of trees, reduction in living space due to population increase, and so forth.

During their evolution, living things have been faced with two alternative policies for achieving **adaptability**. They can remain as generalised in structure and physiology as their genetic make-up will allow, enabling them to colonise a variety of habitats. This is the situation that applies in varying degree to most plants and animals. One of the reasons for the evolutionary success of man has been that, unlike our monkey and ape (anthropoid) relatives, we have remained generalised and therefore highly adaptable. Alternatively, the course of evolution can lead towards **specialisation**, involving a restriction to a narrow set of ecological conditions. Thus, some species of tapeworms are so specialised in their requirements that their distribution is restricted to a single species of host. In such conditions variation will be a disadvantage. It is significant that many cestodes are self-fertilising and they therefore exhibit a high degree of homozygosity. Again, the life cycle of the large blue butterfly, *Maculinea arion*, depends entirely upon an intimate symbiotic association at a critical time in the life cycle between its larvae and certain kinds of ants. If this does not occur the larvae die. Such species, while achieving adaptation to an extremely limited range of conditions, are walking an evolutionary tightrope, for should the populations on which they depend fail, they themselves inevitably face extinction. Once such a course of specialisation has been pursued events cannot be reversed, for in the process, flexibility has been sacrificed and with it the capacity to evolve further.

## Variation and the environment

Although the environment is in a constant state of change, the magnitude of fluctuations and the rate of occurrence of these vary greatly. At one extreme, human activities may perpetrate changes of major proportions at an alarming speed. Typical examples are the felling of woodland, poisoning of the atmosphere with sulphur dioxide, and polluting of fresh water with nitrates and detergent. At the other extreme, stable ecosystems, such as mature forest which has reached the climax of its development, may remain relatively unaltered for a period of many years. In conditions of rapid change, high variability will be an advantage to a species in providing opportunities through natural selection of new and better adapted forms. In conditions of stability the reverse will be true; selection pressure will favour forms that have already proved to have high survival value.

The principle is well illustrated by Darwin in his book *Different Forms of Flowers on Plants of the Same Species* published in 1877. The common primrose, *Primula vulgaris*, is pollinated by insects that are attracted to the nectar secreted by glands at the base of the corolla. Flowers can occur in four different forms (Fig. 5.1) although Darwin only studied the two commonest. The pin-eyed type has a long style with its round stigma projecting through the mouth of the corolla tube like a pin's head (Figs 5.1(a) and 5.2(a)). The five anthers surround the style and are situated about half way down it. The thrum-eyed form has a shorter style and the stigma is invisible from the outside. (The word 'thrum' is borrowed from the

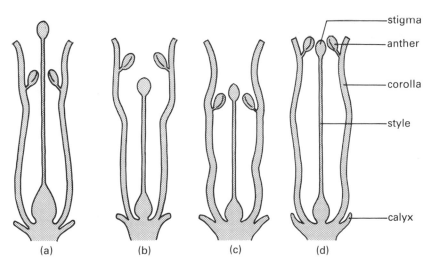

**Fig. 5.1** Heterostyly and homostyly in flower structure in the primrose, *Primula vulgaris*: (a) pin, (b) thrum, (c) short homostyle, (d) long homostyle

**Fig. 5.2**   The common primrose, *Primula vulgaris*: (a) pin-eyed, (b) thrum-eyed

weaving industry and refers to bunches of short, loose threads that are eventually cut away and are known as thrums.) In this kind of flower, the opening of the corolla is surrounded by the ring of anthers (Figs 5.1(b) and 5.2(b)). The two forms are said to exhibit **heterostyly**. On visiting a thrum flower an insect is dusted with pollen as it enters and leaves. If it now visits a different plant, the transfer of pollen to the stigma of a second flower is more likely to be achieved if it is a pin rather than a thrum. Such an arrangement, where a species exists in two different forms, is said to be **polymorphic**. In the primrose, polymorphism is evidently a structural device for promoting cross-pollination and thus maintaining variation.

Darwin carried out numerous experiments involving artificial crosses in primroses and his results are summarised in Table 5.1. The greatly increased fertility of the two normal crosses compared with the abnormal supports the hypothesis that the polymorphic flower structure does, indeed, lead to a high degree of cross-fertilisation under natural conditions.

**Table 5.1**  Darwin's crosses of the primrose, *Primula vulgaris*

| Cross | Flowers pollinated | Capsules set | Average seeds per flower pollinated |
|---|---|---|---|
| *Normal* | | | |
| thrum × pin | 8 | 7 | 56.9 |
| pin × thrum | 12 | 11 | 61.3 |
| *Abnormal* | | | |
| thrum × thrum | 18 | 7 | 7.3 |
| pin × pin | 21 | 14 | 34.8 |

As Ford [5.1] has pointed out, several other characteristics are closely associated with the pin–thrum mechanism. These include such diverse features as the length of the style, the shape of the stigma, the rate of growth of the pollen tube, and the physiological reaction of pin and thrum flowers to different kinds of pollen. Such characteristics could all be under the control of a single pair of alleles having multiple (**pleiotropic**) effects. In fact, experimental evidence shows that they are due to a series of genes located so close together on homologous chromosomes that they are virtually prevented from crossing-over and therefore function as a single gene block. Such a genetic unit is known as a **supergene** (see p. 69). In the example that follows supergenes will be referred to, for simplicity, by single letters as if they were units.

In most primrose populations, pin and thrum plants occur in roughly equal numbers (the two kinds of flowers are never found on the same plant). We now know that a single pair of supergenes ($S$ and $s$) is involved as a control mechanism, thrum being dominant to pin. As we saw from Darwin's results, the abnormal cross of thrum × thrum ($SS$) is almost infertile while pin × pin ($ss$) is also about half as fertile as the normal. In natural conditions most thrum plants will be heterozygous ($Ss$), being the products of thrum × pin crosses. Crossing with a pin flower therefore involves a back-cross as is illustrated in Fig. 5.3.

Pursuing the primrose polymorphism a stage further, we can ask why it is that the normal heterozygotes are so much more fertile than the abnormal homozygotes? The answer is to be found in the differential growth rate of the pollen

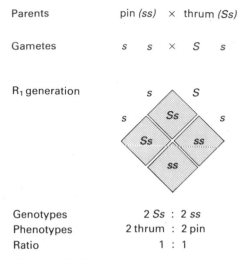

| Parents | pin *(ss)* × thrum *(Ss)* |

| Gametes | *s*   *s*   ×   *S*   *s* |

R₁ generation

| Genotypes | 2 *Ss*  :  2 *ss* |
| Phenotypes | 2 thrum  :  2 pin |
| Ratio | 1  :  1 |

**Fig. 5.3**   Genetic basis for equality in numbers of pin and thrum plants in the primrose, *Primula vulgaris*

tubes. Thus, if a mixture of 'normal' and 'abnormal' pollen is applied to a stigma, the 'normal' always grows the faster and so achieves fertilisation. Another aspect of some significance is the way in which pollen behaviour is inherited. Since pollen grains are haploid, each will contain one member of the pair of alleles controlling flower polymorphism. Those derived from thrum (*Ss*) parents will carry *S* and *s* in equal proportions. But irrespective of their genetic constitution, all are found to behave in a *thrum* way. Similarly, all pollen from *ss* plants behaves as pin.

Two mechanisms thus control compatibility:

(i) **Nuclear**, acting through the genes carried by the pollen grain chromosomes.
(ii) **Non-nuclear** (see also p. 70) or **cytoplasmic**, the behaviour of the pollen being determined by the cytoplasm of the parent.

As we saw earlier, a supergene consists of a block of genes occurring so close together on a chromosome that crossing-over is almost precluded. But on rare occasions it occurs none the less, and this has happened in the primrose, where the genes responsible for the development of the sexual parts of the flower are slightly further apart on the chromosome than the rest. Since the situation has evolutionary implications, it is worth following it further; but in doing so, we need to be a little more precise in our genetic nomenclature. Thus, the dominant thrum condition controlling the female parts of the flower, such as the length of the style and compatibility with different kinds of pollen, can be referred to as *G*; the dominant male parts, such as the position of the anthers, are controlled by gene *A*. The recessive alleles are *g* and *a*. Therefore, *ss* (pin) = (*ga*) (*ga*) and *Ss* (thrum) = (*GA*) (*ga*). Before crossing-over, the only gametes possible were (*GA*) and (*ga*) but after it, two new kinds are formed, namely (*Ga*) and (*gA*). (*Ga*) gametes produce flowers in which the stigma is in the thrum position and the anthers are at the same level and surround it. This is known as **short homostyle**. (*gA*) gives rise to the reverse situation where the style is long and in the pin condition with the anthers surrounding it (Fig. 5.1). This is **long homostyle**. Unlike the heterostyle plants, the homostyles are adapted structurally to be self-pollinated. Moreover, since a single flower possesses the male parts of the thrum and the female parts of pin or the reverse, it is self-fertile. Homostyles are thus adapted both structurally and physiologically for inbreeding with a reduction of variability, in contrast to heterostyles which promote outbreeding and a tendency towards increased variation.

The heterostyle–homostyle system provides an ideal means of adjusting to diverse ecological conditions demanding both stability and change. In this respect, it is significant that a similar polymorphism is known in at least eighteen orders of flowering plants. Ford [5.1] has studied one of a number of colonies of primroses where long homostyle plants are known to occur (near Sparkford, Somerset, south-west England). In some populations their incidence reached a level of 70–85 per cent: in others it was lower at around 15–35 per cent. Fluctuation between the two extremes was a fairly frequent occurrence suggesting a degree of environmental variation. During the 1940s and '50s the species was common, high density being associated with many homostyles. Subsequently, environmental changes occurred, perhaps due to the use of pesticides, and the numbers of primroses declined drastically. So, too, did the proportion of homostyles, to 2–3 per cent. Evidently, the changed conditions now favoured outbreeding rather than inbreeding. This hypothesis was supported in an isolated wood nearby where the plant remained abundant with the proportion of homostyles at a high level, indicating ecological stability in contrast to the changes that had occurred elsewhere.

## Polymorphism

As we saw in the previous chapter, all evolutionary novelty depends upon mutation of some kind which includes any variation that is inherited, other than that resulting from segregation and recombination of genes. Many of the mutations that we observe in nature are of a somewhat drastic kind such as albinism in plants – the inability to produce photosynthetic pigments – or melanism in moths – the production of black forms coloured by the pigment melanin. At their first appearance, such varieties could be expected to occur with a frequency of around one in a million. In the absence of conditions in which their particular characteristics have survival value, they will usually be eliminated either through predation or inability to breed.

But in the example of the primrose just discussed, the situation was different, for here even the rarest variants (homostyles) occurred at an appreciable level. Where two or more forms of the same species exist together with the rarer of them at a frequency above that of recurrent mutation, the situation is known as **genetic polymorphism**. As the primroses showed, polymorphic forms can be sensitive indicators of the action of natural selection; indeed, they themselves are products of it. A comparable situation must have occurred in the many species of moths, such as the peppered moth, *Biston betularia* (p. 97), which produced dark melanic forms that later became established in industrial areas. Their origin will have been in unpolluted surroundings where they would be an easy prey to birds on account of their visibility. But as the environment became more polluted so the level of their survival increased, until it attained a numerical balance with the typical form. While the process was going on, the proportions of the two variants were constantly changing in favour of the new type. This is known as **transient**

**polymorphism** and characterises a state of instability. Eventually, stability was achieved depending upon the relative advantages conferred by the two forms in a particular ecological situation. This is **balanced polymorphism**. To achieve equilibrium, it follows that a balanced polymorphism must pass through a transient stage.

### Polymorphism and natural selection

In Chapter 4, reference was made to a blood condition in man known as sickle-cell anaemia and quantitative data were used to illustrate the application of the Hardy–Weinberg principle (p. 74). The story can now be taken a stage further, since it illustrates the importance of polymorphism in circumstances of powerful selection where the principal selective agent is known.

Human blood can contain various kinds of haemoglobin. At birth, the form still present is largely foetal haemoglobin (HbF) but most of this has disappeared after about 4 months. Adult haemoglobin (HbA) consists of two parts, the greater being $HbA_1$ and the smaller $HbA_2$. A mutation confined largely to Africans, and with no dominance, converts the $HbA_1$ to $HbS_1$, leaving the $HbA_2$ unaltered. We therefore have three haemoglobin genotypes $AA$ (normal), $AS$ (heterozygous), and $SS$ (sickle-cell). The homozygotes from which the condition obtains its name are easily identified as their red blood cells collapse, assuming a characteristic sickle shape (Fig. 5.4) and losing their haemoglobin. This causes anaemia and other problems such as an upset in liver metabolism and blockage of the small blood vessels. Sicklers have a low expectation of life, often dying in childhood and seldom surviving beyond the age of 25.

The heterozygotes have only half their quota of the $HbA_1$ and the full amount of $HbA_2$, the remainder being HbS. In circulation, their red blood corpuscles are indistinguishable from normal but the addition of a reducing agent to a drop of

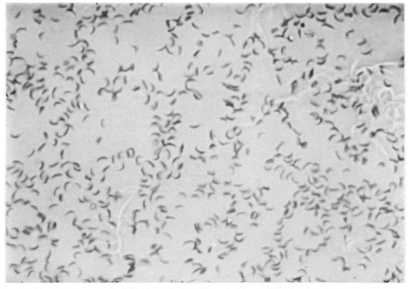

**Fig. 5.4** Blood of an individual carrying the sickle-cell gene in the homozygous state

blood, removing the oxygen, causes the HbS cells to collapse and become sickle-shaped. This is therefore a reliable means of separating *AS* individuals from normal *AA*. In parts of Africa no less than 4 per cent of the children born are *SS*, and *AS* individuals can comprise as much as 40 per cent of the population. At first sight, it seems surprising that a gene exerting such drastic effects can remain at a high frequency in the face of natural selection. How can this have come about?

We now know the answer to this question. It is to be found in the fact that HbS haemoglobin, even in the half amount present in heterozygotes, provides powerful resistance to a dangerous disease – malignant tertian malaria, which is caused by the protozoan parasite, *Plasmodium falciparum*, and is transmitted by mosquitoes. In areas where malaria is prevalent, heterozygous individuals are at an advantage; they will survive whereas those homozygous for normal blood are likely to succumb to malaria and homozygous sicklers will die of anaemia. A gene that is disadvantageous in the homozygous condition but beneficial as a heterozygote is said to confer **heterozygous advantage** (or **heterosis**). This is one of the principal ways in which polymorphisms are evolved and maintained.

An extraordinary situation has recently been investigated in certain Arabian oases where blood tests of the inhabitants have shown that a considerable proportion are homozygous for HbS. Yet they show no adverse symptoms. They live a normal life to a considerable age and the women experience no problems in reproduction. It appears that in these people the ill-effects of the *SS* genotype are offset by the retention into adult life of a high concentration of the HbF that was present in their blood at birth. Through the action of natural selection, such populations have become protected from the adverse effects of sickle-cell while acquiring the benefit of immunity to malaria. Adaptation to life in a malarious environment has thus been carried a step further in Arabia than it has in Africa. A feature of all polymorphisms is that their variation is discontinuous and they present a balanced set of clear-cut phenotypes. Changes in the selective advantage of the different forms in response to environmental fluctuations result in rapid responses to varying selection pressures. This was well demonstrated in populations of the primrose (p. 79). On our hypothesis, the same should be true for sickle-cell anaemia. A means of testing this idea has been provided by the West African negroes who were transported as slaves to North America some 300 years ago. Judged by its rate of decline in recent years, the incidence of heterozygous sickle cell in West Africa at the time of the slaves' departure must have been in the order of 22 per cent; today it is about 10 per cent. In Central America where malaria still persists, the *S* gene occurs at a level of 20 per cent of the population. But in North America there is no malaria and the agent maintaining the polymorphism is therefore absent. The *S* gene is no longer at a selective advantage compared with normal homozygotes and the average incidence of HbS in the heterozygous state is now less than 5 per cent.

# Natural selection and predation

The selective mechanism controlling sickle-cell anaemia just described was a subtle one in that it involved a delicate relationship between a parasite (*Plasmodium*) and its host (man). The frequency of occurrence of the parasite determined the gene frequency of sickle-cell; in malarious areas it was high, in non-malarious areas, low. The incidence of malaria was therefore **density dependent**. As we saw in Chapter 3, Darwin was well aware of the importance of population size in determining the survival of individuals competing for limited essentials such as food. In this process, predation must play an important part. No doubt this was uppermost in the mind of Darwin's contemporary, the poet Alfred Lord Tennyson when he wrote, 'Nature, red in tooth and claw'.

There have been many studies of predators acting as selective agents. Those on the polymorphic land snails of the genus *Cepaea* are included here not only because they illustrate important principles of natural selection, but also because many of them are easily repeatable in different conditions and with the minimum of equipment. Only two species of *Cepaea* occur in Britain, the brown-lipped snail, *C. nemoralis*, and the white-lipped snail, *C. hortensis*. They are usually easy to separate as adults but not as juveniles [5.2]. *C. nemoralis* is the more variable of the two, the colour of its shell being classifiable into three distinct shades – brown, pink, and yellow. The markings on the shell also vary from unmarked forms to those with five dark bands, with a number of intermediates (Fig. 5.5). *C. hortensis* is less variable and therefore easier to study. The colour of the shell is usually yellow and there are either no markings or five distinct dark bands, intermediates being comparatively rare. In both species shell colour and pattern have been shown to be under genetic control. In *C. nemoralis*, colour is determined by a series of multiple alleles in order of dominance: brown, pink, yellow. Thus, yellow is recessive to the other colours. Again, unbanded is dominant to any form of banding. In *C. hortensis*, the situation is somewhat similar; there are effectively only yellow shells and unbanded is dominant to banded.

Studies of *C. nemoralis* by Cain and Sheppard [5.3] in a variety of habitats such

**Fig. 5.5**  Variation in shells of the brown-lipped snail, *Cepaea nemoralis*

as grassland and woods showed that there was a high correlation between yellow shells and green backgrounds such as downland. In darker places ranging from shaded hedgerows to deepest beechwoods, the proportion of yellow gradually declined almost to zero. Again, banding on the shells showed a clear correlation with the environment, the more varied and dark the environment (e.g. in woodland), the greater the proportion of banding.

Cain and Sheppard showed that the main daytime predator on *C. nemoralis* is the song thrush, *Turdus ericetorum*. A large number of snails were collected at random and each marked with a dot of cellulose paint on the ventral side so that it could not be seen by birds. (The outside of the shell is shiny due to the presence of a horny layer, the periostracum. To make the paint adhere, a small patch of this must be scraped away with a file or penknife. For large numbers of shells a simple mechanical device can easily be constructed [5.4].) The marked population was liberated and subsequently sampled at different times. In mid-April the vegetation was predominantly brown, and yellow shells were at a selective disadvantage. By late April, yellow was of neutral survival value relative to the other colours, but by mid-May it was at an advantage. Important corroborative evidence was provided by predated shells. The song thrush is a bird of habit and always cracks snail shells in the same way holding them in its beak by the lip of the opening and cracking the spire against a stone. A typical 'anvil stone' is shown in Fig. 5.6; it will be seen that all the shells have damaged spires. The other main predators on snails are rodents, particularly rats, but they attack by nibbling away the shell from the front so their predation is easily identifiable. Incidentally, being nocturnal feeders, they select their prey at random relative to shell colour and banding.

Evidence from anvil stones supported that obtained from the marked populations, showing the same seasonal fluctuations in predation on yellow and banded shells. Moreover, the pattern of change was found to be similar in two separate

**Fig. 5.6** Thrush anvil stone with shells of the white-lipped snail, *Cepaea hortensis*, and the garden snail, *Helix aspersa*

localities. Thus, as in sickle-cell anaemia considered earlier, a polymorphic character was at a selective advantage in one situation but at a disadvantage in another.

A rather different approach has been adopted in the study of *Cepaea hortensis* in an area of Portland (Dorset, England) [5.5]. Two grassy banks, each some 360 m long running approximately east–west, supported a large population of snails. Over a period of four years, three 45-m stretches were sampled at the same time every year, one at each end and one in the middle. All predated shells were also collected at the foot of the banks where there was an abundance of large stones providing anvils for thrushes. During the period of study, the proportion of banded individuals, both living and predated, remained virtually unchanged. The relevant data are summarised in Table 5.2.

**Table 5.2**   Selective bird predation on the white-lipped snail, *Cepaea hortensis*

| Area | Banded living (%) | Banded predated (%) | Selection against banded (%) |
|------|------------------|--------------------|------------------------------|
| A  Bank 1 (east) | 19.3 | 23.3 | +4.0 |
| B  Bank 1 (middle) | 38.9 | 49.1 | +10.2 |
| C  Bank 1 (west) | 51.3 | 60.8 | +9.5 |
| D  Bank 2 (east) | 15.8 | 26.3 | +10.5 |
| E  Bank 2 (middle) | 30.7 | 40.6 | +9.9 |
| F  Bank 2 (west) | 50.6 | 55.8 | +5.2 |

Two conclusions of particular interest emerge from these results:

(i) Among the living populations, the pattern of banding along the two banks was substantially the same; statistical tests showed no significant difference between them. At the east end, banding averaged about 18 per cent and at the west 51 per cent, with intermediate values between. Evidently, variation in banding along the length of each bank took the form of a **gradient** with a low value at one end and a high value at the other. Such gradients of variation occur quite commonly in wild populations and are known as **clines**. They may be only a few hundred metres long, as in *Cepaea* on Portland, or they may extend many hundreds of miles, as in the wren, *Troglodytes troglodytes*, in Britain. Here, there is a cline in size from large birds in the north to smaller ones further south. The question in *Cepaea* is whether bird predators could be a major controlling factor as we might expect from the findings with *C. nemoralis*. Let us return to the data in Table 5.2.

(ii) The figures for predated shells reveal an unexpected situation. In every instance selection was against banding irrespective of its incidence in the live population. Viewed against their natural background of grass, both banded and unbanded individuals appear equally obvious to the human eye (Fig. 5.7), but how they look to a bird is difficult to say. Clearly, there is no evidence to suggest that the cline of banding could have been the result of differential predation. The data do, however, reveal a well-known fact of bird feeding behaviour, namely the development of a **searching image**. The predator comes to associate a particular feature of its prey (such as banding on a snail shell) with a source of food. Thereafter, attention tends to be confined to this

**Fig. 5.7** Forms of the white-lipped snail, *Cepaea hortensis*, in their natural environment

feature irrespective of its frequency. These findings serve to underline the importance, when conducting investigations of this sort, of appreciating the behaviour patterns of the possible predators concerned.

### Endocyclic selection

When studying the action of natural selection we tend to think of it only in relation to adults. In fact, it can occur at any stage of the life cycle. Applying this idea to *Cepaea* on Portland, a population consisting entirely of *C. hortensis* was studied, the living individuals being divided into two categories – juveniles with shells of 10 mm diameter or less, and the rest. Sampling over three successive years revealed that whereas the older individuals were 19.9 per cent banded, among the juveniles the figure was only 10 per cent. If birds were the selective agents concerned, differential predation among juveniles must have been against *unbanded*; the reverse of the findings for adults. A situation of this kind in which the direction of natural selection changes with the phases of the life cycle of an organism is known as **endocyclic selection**.

### Non-predatory selection

In both species of *Cepaea*, many instances are known where stabilisation of a variable such as banding appears to bear no relation to the colour of the background or predation by birds. Over a considerable area, a particular pattern of variation may predominate for no apparent reason. In *C. nemoralis* this appears to be commonly associated with conditions subject to drought during the summer. Cain and his colleague J. D. Currey have named these localised stabilisations **area effects**. In a polymorphic species, such physiological adjustments to particular ecological requirements could well be the outcome of heterozygous advantage. As we have already seen genes can have multiple effects, some of them structural, others physiological. One way in which these could have developed is through close linkage and the formation of supergenes (p. 69).

The problem of distribution of *C. hortensis* on Portland thus remains unresolved. Although bird predation was extensive, it was evidently not a factor influencing the pattern of banding at any stage of the life cycle. We must therefore look for an explanation in terms of area effects. But in this instance, the environmental factors concerned must have been graded to account for the cline which has proved to be such a stable feature of the population.

## Isolation and evolution

The stability of variation among plant and animal populations is a most striking feature. Indeed, we would expect this to be so in situations controlled by the powerful action of natural selection. The most surprising thing about the cline banding in *C. hortensis* on Portland, and similar situations elsewhere, is that they should occur at all. Marking experiments have shown that, on average, a snail moves about 6.4 m a year (maximum 85 m). In order to maintain the cline, the selective forces operating at each end must have been exceedingly strong to overcome the effects of gene flow.

Taking this idea a step further, if such powerful selection agents exist, what are the circumstances in which a continuous interbreeding population of animals or plants can become split into two or more isolated races that could eventually become new species? There are several possibilities. The more usual situation is when isolation is due to a **geographical barrier** of some kind that prevents further breeding (**allopatric evolution**). Typical physical barriers are stretches of water, forests, and areas of human industrialisation. A second, rarer possibility involves no ecological barrier, only the localised action of strong selective forces on particular variants, resulting in a discontinuous pattern of distribution (**sympatric evolution**).

### Allopatric evolution

When considering the influence of the voyage of the *Beagle* on Darwin's ideas about evolution (p. 41), we encountered a typical example of allopatric evolution in the Galápagos Islands. Here, in conditions of isolation, the different islands support distinct forms of tortoises and birds, mostly quite unlike those on the mainland of South America that form separate breeding groups or **demes**.

But we need not visit the Pacific to observe allopatry; numerous examples are near at hand, for instance in the Isles of Scilly off the extreme south-west coast of England. Here, the effects of isolation are well illustrated in the meadow brown butterfly, *Maniola jurtina*, a common inhabitant of grassland. A feature in the adult that has proved a useful index of variation is the number of spots on the hindwings, which range from 0 to 5 (Figs 5.8 and 5.9). Being under polygenic control, this character responds quickly to the effects of natural selection. In themselves, spots are trivial, although there is now some evidence that they may play a part in courtship. But the gene system controlling them has been shown to influence such diverse characteristics as the length of the life cycle, survival and behaviour patterns in the adult, and susceptibility to pathogens such as bacteria in the larva [5.6].

(a)

(b)

**Fig. 5.8**   The meadow brown butterfly, *Maniola jurtina*: (a) male with wings closed showing the two-spot condition, (b) female with no spots

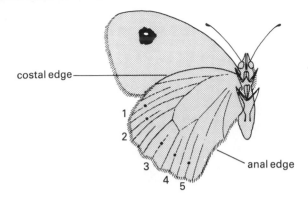

**Fig. 5.9** Positions of spots on the hind wing of the meadow brown butterfly, *Maniola jurtina*

The Isles of Scilly form an archipelago some 48 km to the west-south-west of Land's End. Five large inhabited islands have areas of 250 ha or more. Many of the small islands are mere rocks, the size of the largest being 16 ha (Fig. 5.10). Consistent sampling over a period of 9 years revealed a remarkable stability of spot distribution in both sexes. As the females have proved the more sensitive to environmental changes, attention here will be confined to them. As will be seen from Fig. 5.11, on the three large islands of St Mary's, Tresco, and St Martin's, values of female spotting were nearly equal at 0, 1, and 2 spots, tailing off at 3, 4,

**Fig. 5.10** The Isles of Scilly (smallest islands omitted)

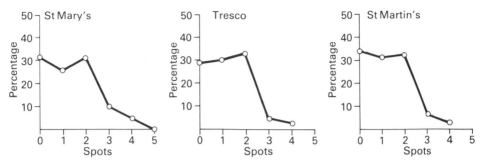

**Fig. 5.11** Spot distribution in female meadow browns on three large islands in the Scilly Isles

and 5. Statistical analysis showed that the individual variations were not significant. In contrast, the small islands with meadow brown populations showed wide but consistent variations in spotting, each being distinct (Fig. 5.12).

The situation on the small islands of Scilly closely parallels that of the Galápagos in that, judged by spot distribution, a different race and breeding group (**deme**) of meadow browns occurs on each. One of the most striking features of the islands is that each possesses its peculiar and easily identifiable form of ecology. The isolated populations of butterflies have thus become adapted to the special conditions prevailing in each of their restricted habitats. But how do we explain the flat-topped spot distributions common to the three large islands? It could be argued that these are merely the result of accumulating a number of different

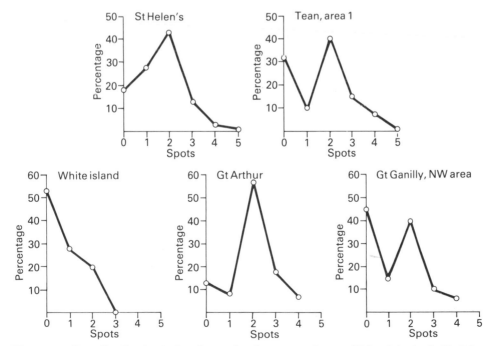

**Fig. 5.12** Spot distribution in female meadow browns on five small islands in the Scilly Isles

samples with diverse patterns. This hypothesis was tested by comparing a number of samples taken from different places; all showed the same distribution. Evidently, the existence on the large islands of a diversity of ecological habitats has enabled the butterflies to adjust to the average of the conditions that prevail there. Like the Galápagos birds, all the spot distributions of the meadow brown populations on the Isles of Scilly show a marked difference from those on the west Cornish mainland from which they were presumably derived, perhaps some 6000 years ago, before the land bridge with England disappeared.

### Sympatric evolution

As an example of sympatric speciation, it could be claimed that the evolution of the cord grass, *Spartina townsendii*, resulting from the hybridisation of two other species and subsequent polyploidy (p. 65), was a typical instance, since all three species colonised the same kind of ecological conditions (mud flats) and no physical barriers existed between them. But this was speciation of a special kind. Here, we are concerned with variants within a single continuous species which are selected differentially resulting in a discontinuous pattern of distribution.

The principle of sympatry is well illustrated in the meadow brown butterfly, *Maniola jurtina*, considered earlier. It will be recalled that values for the number of spots on each hindwing range from 0 to 5 (Fig. 5.9). Extensive studies covering more than 30 years have established that two characteristic spot distributions exist in southern Britain. They occur in both sexes, but since the female is the more variable and therefore shows a greater range of fluctuation, it will be considered here. Samples taken across most of the southern counties from Suffolk westwards have a spot distribution unimodal at 0 (Fig. 5.13(a)). But in east Cornwall, the pattern changes to bimodal with the larger value at 0 and the smaller at 2 spots (Fig. 5.13(b)). We now know that spotting in the meadow brown is controlled by several pairs of alleles with multiple effects. Following the example of *Cepaea* on Portland described earlier (p. 88), we might predict that the transition from one pattern to the other would be gradual and take place over an appreciable distance (**cline**). In fact, a very different situation exists. This can be summarised as follows:

(i) The changeover from a southern English to an east Cornish distribution takes place with surprising abruptness, sometimes in a matter of a few yards. Far from a gradient of variation, the two distributions sometimes become more different as they approach one another (**reverse cline**).

(ii) The boundary area between one form and the other is not related to any physical barrier.

(iii) The position of the boundary has fluctuated during the 20 years it has been known over a distance of about 96 km (60 ml) and with roughly a 10-year periodicity.

(iv) Apart from fluctuations in the position of the boundary, the essential features of the system have never changed. In particular, spotting at the east and west ends of the transect has been stable over the whole period of study.

(v) The situation is not confined to a single transect. Thus, the pattern is the same both north and south of Dartmoor.

(vi) We know that the southern English stabilisation is not peculiar to Britain but extends southeastwards for some 4800 km (3000 ml) into eastern Europe.

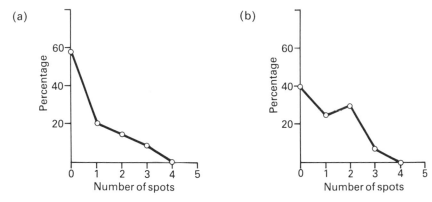

**Fig. 5.13** Spot distribution in females of the meadow brown, *Maniola jurtina*: (a) southern England, (b) east Cornwall

Before we leave this example of sympatry by a species of insect, two other aspects bear consideration. In order that sympatric evolution can occur, powerful selective forces must be at work in the populations concerned, restricting to varying extents the gene flow occurring between them. But what are these forces? All the evidence suggests that predation and other agents acting directly on the adults play only a minor part. Laboratory investigations [5.6] have shown that the gene system controlling spotting also exerts profound effects on the mortality of the larvae, particularly their susceptibility to pathogens such as bacteria. It seems likely, therefore, that some form of **endocyclic selection** (p. 89) exerts a major influence in controlling gene frequencies in *Maniola* populations, but how this operates is still not certain.

## Other mechanisms of isolation

In the previous pages we have seen how a measure of isolation is an important prelude to the formation of new races and, eventually, of distinct species. Geographical barriers promote allopatric evolution and powerful selection pressures are mainly responsible for sympatric changes.

As a rule, species are discrete units and most will not hybridise. If they do, the offspring are usually infertile. Isolation can therefore be **genetic**. It could be argued that the extent of genetic isolation achieved by a particular variety is a measure of the degree of its advancement towards becoming a new species. Among animals, an important aspect of this process is the evolution of sexual behaviour. In general, major variants resulting from mutation such as albino blackbirds and unusually marked butterflies, find greater difficulty in acquiring a mate than do normal individuals. Indeed, in extreme instances, the aberrant individual may be driven away from the colony. Again, the evolution of bright colours in birds (usually the male) and complex behaviour patterns in courtship all have strong selective value for the most successful individuals.

Some closely related species of both plants and animals are capable of hybridisation with the production of fertile young, but this seldom occurs because the respective life cycles are unsynchronised. Among plants this may be because flowers are formed at different times or on account of an incompatibility

**Fig. 5.14** The false oxlip, a hybrid of the primrose, *Primula vulgaris*, and cowslip, *P. veris*

mechanism as occurs in the primrose (p. 79). In insects the two species may not be on the wing together. Thus, the common marbled carpet moth, *Dysstroma truncata*, can pair with the dark marbled carpet, *D. citrata*, and produce fertile offspring, but it seldom does so because the adults of the single brooded *citrata* are on the wing between the two broods of *truncata*. In such circumstances, isolation can be said to be **seasonal**. A change in the rate of either life cycle could result in a new species.

Again, isolation of two interbreeding species may result from their particular environmental requirements resulting in varying degrees of **ecological** isolation. Thus the primrose, *Primula vulgaris*, and cowslip, *P. veris*, hybridise readily to produce the so-called false oxlip (Fig. 5.14). Incidentally, it is worth noting that both species have the same number of chromosomes ($2N = 22$), but since these will not be homologous, the hybrid is unlikely to be fertile. However, should polyploidy occur (see p. 63) fertility would then become possible. The likelihood of crossing is, however, much reduced by the different ecological requirements of the two species, the primrose preferring neutral or slightly acid soils whereas the cowslip occurs on chalk and limestone. This accounts for the comparative rarity of the false oxlip.

## Natural selection and pollution

It is doubtful if any part of the Earth's surface now remains where the ecology has not been affected in some way by human activity. Much of the worst disruption has resulted from industrialisation, which has caused the release of many different pollutants, some of which have had profound effects in promoting the evolution of new forms of plants and animals, and eliminating others.

Among the numerous examples studied, the best documented is that of industrial melanism, a situation in which more than a hundred species of moths, also other insects such as ladybirds, have developed dark-brown or black (**melanic**) populations correlated with the level of pollution of their environment. This has included not only the deposition of solid matter such as soot on vegetation but also the release into the atmosphere of noxious gases such as sulphur dioxide. Such gases are poisonous to the lichens that colonise walls and the trunks of trees. (Lichens provide an important background against which resting insects are camouflaged from predation by birds.) Outstanding among the studies on melanism have been those of H. B. D. Kettlewell on the peppered moth, *Biston betularia*, which are fully covered in most general textbooks as well as in more specialised publications [5.7, 5.8, 5.9]. This work is worth consulting not only as an example of natural selection in a polluted environment, but also as an illustration of good experimental design.

A. D. Bradshaw and his associates have studied a different aspect of pollution – that associated with the mines of such metals as lead, tin, and zinc [5.1, 5.9]. Such localities are not difficult to find and are usually accessible for study. They are identifiable by the lack of vegetation which gradually declines as the site of the mine is approached. Yet, on the mine workings themselves there are frequently scattered clumps of plants such as the grasses, *Agrostis tenuis* and *Festuca ovina* (Fig. 5.15). Occasionally, as in the Mendips (Somerset), there are also unexpected colonists such as the sea campion, *Silene maritima*. How do some species manage to withstand pollution while others apparently cannot?

**Fig. 5.15** Discontinuous clumps of heavy metal-tolerant species of grass colonising the area of a lead and zinc mine

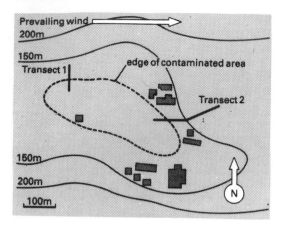

**Fig. 5.16** Plan of the mine studied by McNeilly

McNeilly [5.10] made a detailed study of metal tolerance in the grass *Agrostis tenuis*, growing on the site of a copper mine in Wales. Tests showed the soil to be so contaminated that plants collected from elsewhere and planted in it soon died. An **index of tolerance** was devised to provide a basis for quantification. This was based on the extent of root growth in different concentrations of copper in a standard time under carefully controlled conditions. The investigation involved two transects (Fig. 5.16) of the contaminated area – one at the west end across the prevailing wind; the other further east along the line of the wind. McNeilly's results are summarised in Fig. 5.17 and his main findings are as follows:

(i) In the first transect (across wind), the change of tolerance between toxic and non-toxic soils was abrupt and took place in a metre or so.
(ii) In the second transect (downwind), tolerance was discontinuous and gradually faded out over a distance of 150 m of relatively uncontaminated ground.

Several deductions can be made from these results:

(i) The direction of the wind clearly played an important part in influencing the distribution of *Agrostis*. This must have occurred through the scattering of pollen carrying the genes that controlled tolerance.
(ii) In the first transect, the abrupt change between mine and non-mine populations recalls situations encountered in *Maniola* (p. 90) and elsewhere, and indicates the action of powerful selective forces influencing the survival of plants that possess a degree of tolerance. This view is supported by evidence from the second transect, showing that once selection was relaxed in the relatively uncontaminated ground, the survival value of metal tolerance was less important.
(iii) The difference in mean tolerance between the adult and seed populations is significant for this suggests selection during the life of the young plants.

When a particular variation conferring increased survival is of the discontinuous kind we could expect changes of gene frequency in a population to be detectable in just a few generations. This is what we have found in our studies of spot distribution in the meadow brown butterfly. In a mine, once soil pollution has occurred and selection pressure in favour of tolerance is therefore high, we might

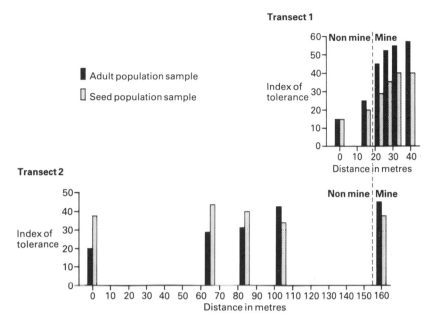

**Fig. 5.17** Copper tolerance in populations of the grass, *Agrostis tenuis*, in the vicinity of a Welsh mine (after McNeilly 1968)

expect some variable plant species to achieve colonisation in a relatively short period of time. Bradshaw has obtained evidence of this from grasses such as *Agrostis* and *Festuca* growing under newly erected galvanised wire fences covered with a protective coating of zinc. Evidently, an appreciable level of tolerance can be evolved in less than 30 years.

Finally, it should be noted that the study of pollution tolerance among plants has been by no means only an academic exercise. The discovery of species capable of withstanding different metals in high concentration has had great practical significance, for it has enabled previously derelict regions such as mine workings and slag heaps to be replanted. In this way they have been converted into areas of vegetation with both aesthetic and amenity value.

## Continuous variation and natural selection

So far, we have been mainly concerned with a kind of variation where there are fairly clear-cut distinctions between the different forms (**discontinuous variation**). Polymorphic characteristics provide typical examples. But much of the variation we see around us is not of this kind. Features of plants such as the weight of seeds and the extent of growth, also size and colour in animals, frequently exist as graded series. When plotted as a frequency histogram they produce a bell-shaped curve of **normal distribution** (Fig. 5.18). Although at first sight such species do not appear to conform to Mendelian laws of inheritance, their peculiar form of variation is usually explicable in terms of multiple alleles or polygenes (p. 62) and is known as **continuous variation**.

Figure 5.18 illustrates three ways in which natural selection may act on a variant within a population with continuous distribution. One of the commonest

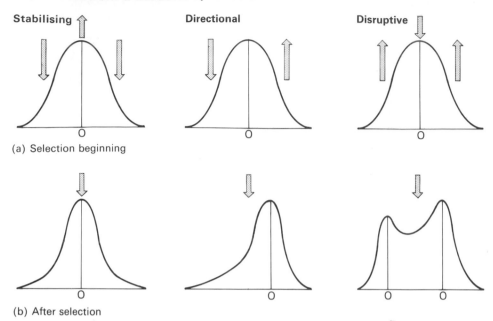

**Fig. 5.18** Three kinds of natural selection: (a) selection beginning, (b) after selection. O = optimum value. Arrows pointing upwards indicate selective advantage, downwards show disadvantage

situations is when the optimum value corresponds to the mean or nearly so, the two extremes being disadvantageous (**stabilising selection**). The widely quoted study by M. N. Karn and L. S. Penrose between 1935 and 1946 on the relationship between human birthweight and mortality provides an example. They collected the birth weights of some 3760 children born in a London hospital over a period of 12 years, also data on the survival of babies. The relationship between mortality and weight is shown in Fig. 5.19, deaths being gauged by the percentage of children failing to survive to 4 weeks. It will be seen that the optimum birth weight (O) is close to its mean (M). On either side, the expectation of survival with increasing or decreasing weight declines rapidly and reaches a minimum at the two extremes.

It has been estimated that as much as half the variation in human birth weight may be genetic, the remainder being the result of environmental factors such as the nutritional status of the mother. One outcome of stabilising selection is to reduce the spread of variation about the mean; indeed, with the powerful elimination of genes having disadvantageous effects, we might expect to find the inherited component reduced to negligible proportions in a matter of a few generations. Yet, there is no evidence that such a change is happening, which raises the question as to how losses in variation are made good. The answer is to be found in the store of variability made available through recombination (p. 58) which must replace elements of the gene pool as fast as they are eliminated. Recurrent mutation will also contribute to the fund of new variants.

In a variable species, a form may be at an advantage in a particular environment and achieve predominance. The frequency distribution will no longer be normal but will become skewed in the direction of the successful variant. This is known as

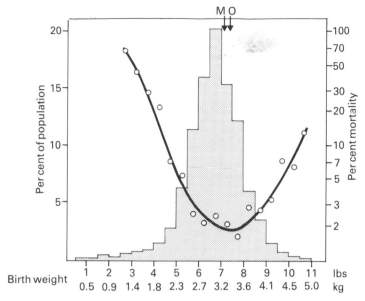

**Fig. 5.19**  Relationship between child mortality and birth weight illustrating the action of stabilising selection. M = mean birth weight; O = optimum birth weight (after Mather 1966)

**directional selection** (Fig. 5.18). A common night-flying moth the dark arches, *Xylophasia monoglypha*, occurs in a variety of colours ranging from buff to almost black. In southern England the commonest colour is medium brown (the mean) and selection therefore appears to be stabilising, or nearly so. But in the north of Scotland there is strong selection in favour of black, the other colours being considerably rarer. Selection here is of the directional kind. Melanism in *X. monoglypha* is unrelated to industrialisation as the black form attains its maximum in some of the most pollution-free areas of Britain. Evidently, blackening somehow aids survival, perhaps by absorbing more heat or being linked genetically to other genes conferring physiological advantages in a rigorous environment.

A third pattern of selection can occur if a variant is at an advantage at the two extremes but at a relative disadvantage in the area of the mean. This is **disruptive selection**, as is illustrated in Fig. 5.18. Attempts to simulate this under laboratory conditions using insects such as fruitflies (*Drosophila*) and houseflies (*Musca*) have met with variable success although some seem to have succeeded. In wild populations the model provides an explanation of the existence of clines such as occurred in *Cepaea hortensis* on Portland (p. 88).

## Coevolution in plants and animals

The existence of ecosystems necessitates close relationships between the plants and animals that compose them. It is therefore not surprising to find that in the course of evolving such relationships the organisms themselves have reacted with one another producing strong reciprocal links. Such evolutionary interaction is known as **coevolution**. Many of these relationships centre round the production of poisonous substances by plants and the varying ability of different animals to detoxify them or to use them for other purposes. Thus, H. W. Kircher and his associates studied the relationships between different species of the fruitfly, *Drosophila*, and the cactus, *Lophocereus schottii*, which produces toxic alkaloids. These were found to be lethal to eight species of *Drosophila*, but not to *D. pachea*, which, having evolved a detoxifying agent, was able to use the rotting stems of the cactus as a breeding place.

A somewhat comparable situation occurs in Britain and is particularly suitable for study. The capacity to produce the highly toxic substance, hydrogen cyanide (**cyanogenesis**), has been evolved independently many times in the plant kingdom, predominantly among flowering plants but also in some ferns and basidiomycete fungi. The precursors of HCN are various cyanogenic glucosides such as lotaustralin which, in the presence of the enzyme $\beta$-glucosidase, are broken down to ketones and the gas liberated. Thus:

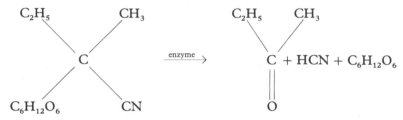

The situation is well illustrated in the bird'sfoot trefoil, *Lotus corniculatus* (Papilionaceae), a common colonist of downland (Fig. 5.20). D. A. Jones and

**Fig. 5.20** Bird'sfoot trefoil, *Lotus corniculatus*, a common colonist of grassland which exhibits cyanogenesis

others have shown that cyanogenesis is under the control of two pairs of unlinked alleles, the dominant *Ac* determining the presence of glucoside and the recessive *ac* its absence. Similarly, the presence and absence of enzyme are governed by the dominant *Li* and recessive *li* respectively. The possible genotypes and their phenotypic effects are summarised in Table 5.3.

**Table 5.3** Genotypes and their phenotypic effects on cyanogenesis in the bird'sfoot trefoil, *Lotus corniculatus*

| Genotype | Degree of cyanogenesis |
| --- | --- |
| (1) *Ac–Li–* | Rapid and strong |
| (2) *Ac–li li* | Slow and weak |
| (3) *ac ac Li–* | None, unless glucoside added |
| (4) *ac ac li li* | None, even when mixed with enzyme or glucoside |

Cyanogenesis in *Lotus corniculatus* (also in the white clover, *Trifolium repens*) is thus a double polymorphism, although it will be noted that some cyanide is produced in the absence of enzyme (Table 5.3(2)), but only when the plants are bruised. The incidence of cyanogenesis varies greatly from place to place ranging from 5 to 80 per cent of a population. Phenotypes carrying the *Ac* gene are easily detectable in the field by testing for HCN with sodium picrate paper [5.11].

The adaptive significance for the plant of this polymorphism has been assessed both in the field and the laboratory, from which it is clear that the production of cyanide when the tissues are damaged provides effective protection from browsing molluscs, particularly the slug, *Agriolimax reticulatus*. However, both cyanogenic and acyanogenic plants are eaten by certain insects, notably the larvae of the six-spot burnet moth, *Zygaena filipendulae*, and the common blue butterfly, *Polyommatus icarus*. The burnet larva has been shown to contain cyanogenic substances at all stages of its life cycle even if it has been eating only acyanogenic plants. The adult moth is day flying and warningly coloured with red hindwings and black and red forewings (**aposematic**), so presumably the cyanide is produced as a deterrent to predators. A hymenopterous parasite of its larva (ichneumon) is *Apanteles zygaenarum*, and this is able to survive inside its host because it produces a detoxifying enzyme which converts cyanide to harmless thiocyanate. It is significant that the enzyme is absent from the nearly related *A. tetricus*. This ichneumon parasitises the larva of the meadow brown butterfly, *Maniola jurtina*, which feeds on grasses where there are no cyanogenic forms. A similar detoxifying mechanism is used by the larva of the common blue, which is free of cyanide even when feeding on toxic plants.

It is now established that many species of molluscs are deterred from feeding on cyanogenic plants both of *Lotus* and *Trifolium*. But there are also selective advantages associated with the *ac li* genotypes. Thus, H. Daday demonstrated a close correlation with the January isotherm and altitude suggesting that acyanogenic plants are better able to withstand cold conditions. In the Alps, an increase in altitude of little more than 1000 m resulted in a decrease in the frequency of the *Ac Li* genotype from about 80 per cent of the *Trifolium* population to zero. The genetic mechanism controlling temperature tolerance is uncertain but it seems likely to be due to genes closely linked to *Ac* and/or *Li* resulting in the formation of supergenes (p. 69). Coevolution of *Lotus* and *Trifolium* with herbivorous animals and their parasites has thus resulted in a balanced polymorphism involving biotic factors on the one hand and climatic factors on the other. There is evidence that the situation, particularly on the climatic side, may be more complex than this account suggests, but a more final conclusion must await further research.

## Evolution by chance?

When considering the genetics of populations (p. 74) we saw that for the operation of the Hardy–Weinberg principle, certain conditions must be satisfied. There must be no mutation, selection or migration, survival must be equal, and reproduction at random. This last requirement presupposes that the genes carried by the gametes pair in the proportions in which they occur in the population. The idea can be illustrated in human families where the XY ($\male$) and XX ($\female$) sex-determining mechanism involves a back-cross, thus ensuring equality in the next generation. In a family with one girl, the probability of a second child being a boy or a girl is $\frac{1}{2}$. But the chance of a second girl following the first is $\frac{1}{2}^2$ ($=\frac{1}{4}$). Similarly the chance of a third girl is $\frac{1}{2}^3$ ($=\frac{1}{8}$) and a fourth $\frac{1}{2}^4$ ($=\frac{1}{16}$). From our own experience (i.e. small sample) we may be able to recall one or more four-girl families but perhaps none with four boys (or the reverse). But in a large sample, such as the population of Britain, four-girl families will be balanced by a roughly equal number of four-boy ones.

Taking this idea a stage further, suppose two alleles, $A$ and $a$, occur in equal numbers in three populations of sizes 500 000, 50 and 20 respectively. What deviations from gene equality would we expect among the fertilisations occurring in the three gene pools? The proportions of genes will be the same as that of the gametes carrying them, which will be double the number of individuals (i.e. 1 000 000, 100 and 40 respectively).

The expected deviation from 50 per cent is determined by calculating the **standard error**. For the large population this is:

$$\sqrt{\left( \frac{500\,000 \times 500\,000}{1\,000\,000} \right)} = \pm\,500$$

Thus, the proportions of $A$ and $a$ genes lie within the range 500 000 ± 500 = between 500 500 and 499 500 (an error of 0.1 per cent). For the population of 50, the standard error is 5 and the proportion of $A$ and $a$ is 50 ± 5 = between 55 and 45 (an error of 10 per cent). For the population of 20, the standard error is 2 and the proportion is 20 ± 2 = between 22 and 18 (an error of 20 per cent).

As the size of a population decreases, so the deviation from expectation increases rapidly. Such changes in gene frequency due to the random effects of gene assortment are known as **genetic drift**. As the previous examples showed, drift is a mathematical certainty and bound to occur in all populations; but its effects will be most evident when numbers are small. Two problems follow:

(i) What are the most likely circumstances in which small populations can occur and be influenced appreciably by drift?
(ii) To what extent is it possible to identify the respective contributions of drift and natural selection, since the two will invariably operate together?

One of the axioms of ecology is that all populations fluctuate in numbers. Such fluctuations are particularly characteristic of animals, where densities may range from maxima determined by the availability of resources to minima, where sexual reproduction becomes uncertain due to the dispersion of the two sexes. At the lowest fluctuation densities, we would expect random gene assortment to become increasingly important and this is sometimes referred to as **intermittent drift**. Suppose that a few migrants from an island colony were to establish the species for the first time on a neighbouring island, they could carry with them a gene frequency very different from the average of the parent population. Thus a new race could be established. Such a process has been called by Ernst Mayr the **founder principle**, and it may well have played a significant part in the formation of some isolated populations.

It has been possible to test this hypothesis on one occasion in the Isles of Scilly using a population of the meadow brown butterfly, *Maniola jurtina* [5.6]. In a small isolated colony on the island of Tresco, female spotting (see p. 92) was found to be quite distinct from that of the main island population. The numbers of the small population were estimated by the process of mark, release, and recapture [5.4] and found to be between 100 and 150 insects. The following year the colony was decimated by a prolonged drought and barely survived. Precise numbers were impossible to estimate but they could not have exceeded twenty. In the following and subsequent years the environment returned to normal and density increased. The spot distribution, instead of assuming some new level and fluctuating as we would have expected under the predominant influence of drift, returned to its former value and remained stable thereafter. On this occasion at least, any effects of drift were masked by those of natural selection.

Situations such as this serve to highlight a potentially important problem in the conservation of rare, endangered species. Provided population numbers can attain a level where natural selection exerts a predominant effect, variation will tend to be adaptive and the population is likely to survive. But as numbers decline, the effects of drift will be correspondingly increased, until a point is reached at which the action of natural selection is so diminished that variation assumes an increasingly random pattern, some of it no doubt being non-adaptive. This could be the prelude to eventual extinction.

# Summary

1   The species is the basic unit within which evolutionary changes can be detected and studied.
2   The environment where evolution occurs consists of physical (abiotic) and living (biotic) components. In adapting to a particular set of environmental conditions, an organism needs to remain as generalised as possible and to avoid excessive specialisation.
3   Heterostyly in the primrose is of evolutionary value as a mechanism for controlling either variability (outbreeding) or stability (inbreeding).
4   Genetic polymorphism provides a sensitive mechanism for promoting evolutionary change. Sickle-cell anaemia represents a balanced polymorphism, the balance of advantage against disadvantage changing with circumstances.

5   In a polymorphic species, predation can be an important factor controlling change. This is well illustrated in snails of the genus *Cepaea*.

6   Although some selection occurs in the adult phase, much of it also takes place earlier in the life cycle and sometimes in a different direction. This is known as endocyclic selection.

7   Area effects in species such as snails indicate that structural differences may be associated with physiological characteristics having survival value. These may transcend the effects of predation.

8   Allopatric speciation occurs when a population becomes divided by physical barriers such as water and forests. Islands such as the Galápagos and Scillies provide typical examples.

9   Sympatric speciation occurs in the absence of physical barriers as is illustrated by the meadow brown butterfly.

10  Other mechanisms of isolation include genetic differences, seasonal factors and variation in ecological requirements.

11  Pollution of the environment by man has provided a powerful influence promoting evolution in wild populations. Heavy metal tolerance in grasses is a typical example with important practical implications.

12  Much of the variation occurring in plants and animals is continuous and provides graded series. Selection can be stabilising (converging towards the mean); directional (producing a skew distribution); or disruptive (favouring the two extremes). Examples of all three processes occur in natural populations.

13  In ecosystems, plants and animals are intimately related and they have sometimes evolved together (coevolution). This has occurred in species of plants with cyanogenic forms in relation to herbivorous animals and their parasites.

14  Genetic drift due to random pairing of gametes occurs in all populations. Its effect is greatest in small colonies, as on islands where it may have played a part in establishing new races. Experimental evidence is uncertain but in at least one instance the effects of powerful selection swamped those of drift.

## Topics for discussion

1   In spite of the severity of natural selection many serious hereditary human diseases still persist. How can we account for this?

2   Why is the importance of genetic drift on the evolution of human populations likely to be less now than it was about 100 000 years ago?

3   To what extent are fluctuations in the numbers of populations likely to influence the rate of evolutionary change?

4   What evidence is there that genetic polymorphism is a powerful factor influencing the evolution of species?

5   What are the advantages and disadvantages of becoming specialised?

6   It has been claimed that the study of natural selection is an academic exercise of no practical value. Is such a claim justified?

7   In what circumstances would be expect to find plants and animals coevolving?

8   To what extent could the founder principle explain the colonisation of the different Galápagos Islands by distinct populations of finches?

# References

**5.1** Ford, E. B. (1981) *Taking Genetics into the Countryside.* Weidenfeld and Nicolson

**5.2** Kerney, M. P. and Cameron, R. A. D. (1979) *A Field Guide to the Land Snails of Britain and North-West Europe.* Collins

**5.3** Cain, A. J. and Sheppard, P. M. (1950) 'Selection in the polymorphic land snail *Cepaea nemoralis*'. *Heredity* **4**, 274–94

**5.4** Dowdeswell, W. H. (1983) *Ecology – principles and practice.* Heinemann

**5.5** Day, J. C. L. and Dowdeswell, W. H. (1968) 'Natural selection in *Cepaea* on Portland Bill'. *Heredity* **23**, 169–88

**5.6** Dowdeswell, W. H. (1981) *The Life of the Meadow Brown.* Heinemann

**5.7** Edwards, G. J. R. (1977) *Evolution in Modern Biology* (Studies in Biology No. 87). Edward Arnold

**5.8** Dowdeswell, W. H. (1975) *The Mechanism of Evolution*, 4th edn. Heinemann

**5.9** Ford, E. B. (1975) *Ecological Genetics*, 4th edn. Chapman and Hall

**5.10** McNeilly, T. S. (1968) 'Evolution in closely adjacent plant populations'. *Heredity* **23**, 99–108

**5.11** Nuffield Advanced Biology (1971) *Laboratory Book.* Penguin

# 6

# Evolution above the species level

In the previous chapter, the process of natural selection leading to evolutionary change was considered mainly below the species level. This is sometimes referred to as **microevolution**. In contrast, the evolution of taxonomic groups at species level and above is known as **macroevolution**. Although such a division means relatively little in terms of the mechanism by which change has come about, it provides a convenient way of categorising the outcomes as we find them today.

But before considering macroevolution and its implications, it will be worth-while summarising the main evidence available that it has actually taken place.

## Evidence for macroevolution

Some of the evidence for evolutionary change in the past has already been considered in connection with the origin of species; here it will be viewed in a broader context and further evidence added.

### Fossil evidence

As we have already seen (p. 19) the fossil record provides the only tangible evidence that evolution has occurred. In spite of its incompleteness, it has none the less provided us with a broad outline of the sequence of plant and animal evolution. Darwin considered evolution to be a continuous process and therefore postulated that the sequence of fossils would exhibit gradual change. However, we now know that changes have taken place over time at greatly differing speeds. Sometimes these changes have, indeed, been continuous, but at other times periods of comparative stability have alternated with those of rapid alteration. How we can account for continuity and discontinuity in terms of evolutionary theory will be considered further in Chapter 8.

### Adaptive radiation

At the levels of the genus and species, we have already encountered a typical example of adaptive radiation among the finches of the Galápagos Islands, which so impressed Darwin (p. 45). Thus, the beaks of the different species are adapted for diets associated with the particular ecological niches they occupy. They range from the formidable bills of seed eaters and the curved beaks of flower-feeders, to the delicate, slender beaks of small insectivores. Although we can never be certain, the supposition here is that this variety of well-adapted species has been derived from a few more generalised ancestral colonists.

At a higher classificatory level, fossil evidence suggests that the first primitive placental mammals appeared at the beginning of the Eocene epoch. Among these

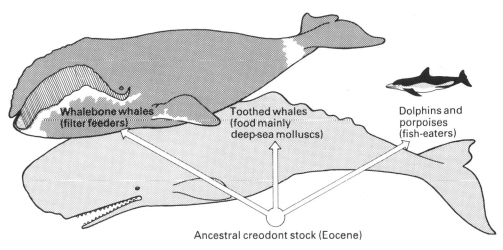

**Fig. 6.1** Adaptive radiation among some marine mammals

was an order (the Creodonta) which diverged into a number of families of carnivores, among them the marine mammals (Fig. 6.1). These include the great whalebone whales with their highly specialised filter-feeding device, which enables them to strain off masses of small crustaceans (krill) from the sea water. Some of these whales attain a weight of 20 tonnes. The toothed whales, such as the sperm whale, feed largely on deep-sea molluscs, while the porpoises and dolphins are specialised for fish-eating.

### Effects of isolation

The animals of the Galápagos Islands also provide us with powerful evidence of the effects of isolation in promoting evolution. Thus, the giant tortoises (*Geochelone*) have developed subspecies characteristic of each island and identifiable by the shape, colour, and thickness of the shell, as well as other bodily features such as size (p. 41). On a worldwide scale, one of the most striking examples of evolution in isolation is provided by the marsupial mammals (those having a pouch or marsupium in which the young develop). During the Mesozoic era when the mammals were evolving, the Earth was still almost a single land mass (Fig. 6.2), with South America joined to Africa and connected with Australia via Antarctica; the bridge with North Africa was not yet severed. The subsequent decline of the marsupials can be accounted for largely through unsuccessful competition with the increasing numbers of placental mammals (mammals where young are attached to the mother by a placenta). Where the two came into conflict for the same ecological niches, the marsupials invariably succumbed. But the subsequent separation of the American and Australian continents afforded the marsupials protection from competition and culminated in their present-day restriction to these two areas. This is particularly true of Australia where the placentals never penetrated (species such as dogs and rabbits are later introductions by man). In the absence of competitors, the marsupials diverged into a number of different forms adapted to a variety of ecological conditions, some of them showing a striking resemblance to their placental counterparts elsewhere (Fig. 6.3).

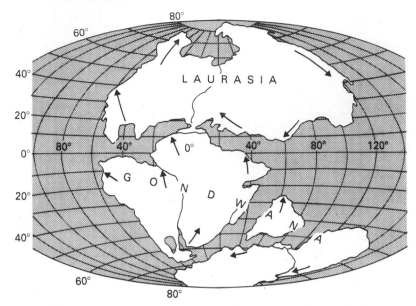

**Fig. 6.2** The Earth as it probably was in the Mesozoic era. Arrows indicate the direction of movement of the continents

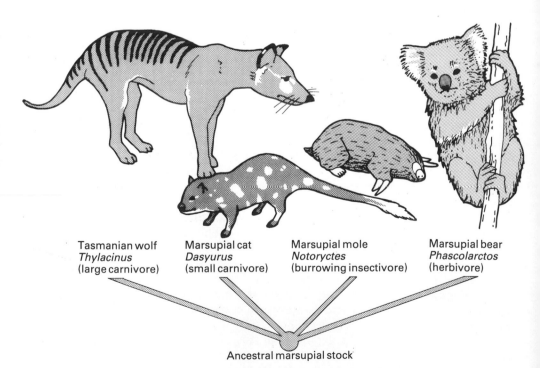

| Tasmanian wolf | Marsupial cat | Marsupial mole | Marsupial bear |
| *Thylacinus* | *Dasyurus* | *Notoryctes* | *Phascolarctos* |
| (large carnivore) | (small carnivore) | (burrowing insectivore) | (herbivore) |

Ancestral marsupial stock

**Fig. 6.3** Divergent evolution among some Australian marsupial mammals in isolation from competition with placentals. Many of the adaptations to different ecological niches show a strong resemblance to those of placentals elsewhere

usually in the form of non-functional stamens. Thus, the figwort, *Scrophularia nodosa*, bears evidence of a former five-rayed symmetry but is now bilaterally symmetrical. Of its five former stamens, four are grouped in two pairs and the fifth is vestigial, never producing any pollen.

*Neoteny*

The form eventually attained by the adults of a species is controlled genetically and can be subject to natural selection during the process of growth. The genes involved determine both the nature of the adult characters and the rate at which they appear, including the onset of sexual maturity. As we have seen, the relationship between the two can vary, and the development of a structure can be so retarded that by the time the adult stage is reached it has hardly appeared at all. This accounts for the existence of vestigial structures discussed in the previous section. In the same way, the speeding up of the development of sexual maturity relative to bodily specialisations can result in an adult organism retaining juvenile characteristics. Such a situation is known as **neoteny** – the precocious attainment of sexual maturity. A classic example of the influence of neoteny on evolution was discovered by Walter Garstang and others who showed that the chordates (animals with a notochord, and the ancestors of vertebrates) are probably derived from the sexually mature larvae of echinoderms such as sea urchins.

The literature on animal evolution contains many instances where the onset of neoteny has played an important part in evolution. One of the most striking has been recorded by Campbell [6.2], and concerns man himself. Thus, the limited body hair of modern man, *Homo sapiens*, and the fair skin of the Caucasoid races may well be the result of the prematurely arrested development of the hair follicles and the melanin pigment cells associated with them. Again, the relatively large brain and absence of brow ridges are characteristic of the *young* stages of apes; as is the rate of growth of the brain compared with that of the rest of the body and the enormous size of the cerebral hemispheres at birth in comparison with that of neighbouring organs.

## Geographical distribution

As we have seen, different species of plants and animals such as marsupial mammals exhibit characteristic patterns of geographical distribution, occurring in some places and not in others. One of the most powerful factors determining species distribution is isolation due to physical or ecological barriers, or both. This has been responsible for the evolution of **endemic species** whose distribution is peculiar to a particular area. The number of endemic species present in a locality will depend largely on how effectively and for how long the area of colonisation has been isolated. Some typical data for flowering plants have been summarised by Tivy [6.3] and are reproduced in Table 6.1.

The highest level of endemism is to be found in oceanic islands with extreme climatic conditions, and in areas of extensive mountain barriers. Continental islands within easy reach of the mainland and only recently separated from it, such as the British Isles, are low in endemic plant species as are non-isolated areas of continents, such as California. Endemism occurs among animals in a similar way. Here, then, is additional strong evidence of past evolution.

Just as the geographical distribution of the larger plants and animals exhibits evolutionary divergence so, too, does that of their respective parasites. As we saw

**Table 6.1**   Geographical distribution of some endemic species of flowering plants (after Tivy 1982)

| Areas | Number of species | Per cent endemic |
|---|---|---|
| *Continental areas* (North America) | | |
| California | 5529 | 38.0 |
| West Virginia | 2040 | 0 |
| *Continental islands* | | |
| British Isles | 1666 | 0 |
| *Oceanic islands* | | |
| Canary | 826 | 53.3 |
| Galápagos | 386 | 40.9 |
| Hawaii | 1721 | 94.4 |
| Juan Fernandez | 146 | 66.7 |
| St Helena | 45 | 88.9 |

earlier (p. 109), during the Mesozoic, South America and Africa were joined together in a single land mass, which connected with Australia via Antarctica. Subsequent separation of the three continents resulted in the isolation of groups such as the flightless birds (ratites). These gave rise in Africa to the ostriches and in South America to the rheas. The closeness of the common origin of ratites is indicated by the similarity of the parasites of these birds today. This extends to their tapeworms (a species of *Houttuynia*), nematodes (sclerostomes) and feather-mites. The evolutionary remoteness of the Australian ratites (cassowaries and emus) is supported by the fact that their tapeworms belong to a different genus (*Railletina*) and their nematodes and feather-mites are also quite distinct.

## Physiological evidence

As might be expected, physiological processes reveal extensive evidence of the evolutionary past of living organisms. This evidence has been studied more extensively in animals than in plants, one of the most striking examples being provided by the blood groups that occur in man and other primate mammals, particularly the human ABO series [6.4]. This well-known polymorphism achieves remarkably varying equilibria in different parts of the world, as is illustrated in Table 6.2.

So stabilised are these values that they have been used as valid criteria for judging the existence of human races. Not only do they differ from one country to another, but they also exhibit distinct evolutionary patterns. Thus, the percentage frequency of group B decreases from east to west: in India it is 37.2, in Iran 22.2, in Germany 14.1, while in France it drops to 6.0. Blood-group distributions also serve to highlight the distinctness of isolated races such as the Hungarian gypsies who differ markedly from other Hungarians, particularly in respect of A and B (Table 6.2). It is still far from clear what selective advantages and disadvantages derive from particular blood groups or combinations of them. At one time it was thought that there is a positive correlation between group O individuals and the incidence of duodenal ulcers. But more critical examination by Clarke [6.5], who used families showing segregation for ulcers as well as containing individuals with

**Table 6.2** Percentage frequencies of ABO blood groups in different races (arranged in increasing frequency of group B) (after Ford 1973)

| Race | Total examined | Frequency (per cent) of blood groups O | A | B | AB |
|---|---|---|---|---|---|
| Australian (aboriginals) | 603 | 54.3 | 40.9 | 3.8 | 1.0 |
| Dutch | 14 483 | 46.3 | 42.1 | 8.5 | 3.1 |
| English (southern) | 3 449 | 43.5 | 44.7 | 8.6 | 3.2 |
| Dutch Jews | 705 | 42.0 | 39.4 | 13.4 | 4.5 |
| Russian Jews | 1 475 | 36.6 | 41.7 | 15.5 | 6.1 |
| Bushmen | 336 | 83.0 | — | 17.0 | — |
| Hungarians | 1 041 | 29.9 | 45.2 | 17.0 | 7.9 |
| Arabs | 2 917 | 44.0 | 33.0 | 17.7 | 4.1 |
| Japanese | 24 672 | 31.1 | 36.7 | 22.7 | 9.5 |
| Russians | 57 122 | 32.9 | 35.6 | 23.2 | 8.1 |
| Negroes (Congo) | 500 | 45.6 | 22.2 | 24.2 | 8.0 |
| Chinese (Canton) | 500 | 45.5 | 22.6 | 25.0 | 6.1 |
| Hungarian gypsies | 925 | 28.5 | 26.6 | 35.3 | 9.6 |
| Hindus | 2 357 | 30.2 | 24.5 | 37.2 | 8.1 |

different ABO groups, showed no evidence of such a relationship. There seems to be some justification on statistical grounds for the assumption that individuals in blood groups A and B are more likely to catch smallpox when it is prevalent, than groups B and O. This has prompted the idea that the present world distribution of ABO may have been influenced by the great pandemics of infectious diseases of former times, particularly those causing a high mortality, such as plague and smallpox.

## Taxonomy: the classification of organisms

Ever since we began to study the plants and animals around us we have attempted to bring order into diversity by assigning them to groups of similar kinds – **taxa**. This process is known as **taxonomy**. The first taxonomists of any note were the Greek philosophers Aristotle (384–322 BC), whose interest was primarily in animals, and his pupil Theophrastus (380–287 BC), who specialised in plants. They believed in spontaneous creation – that living organisms ultimately arose from non-living matter. The idea that one form could ever change into another did not occur to them: all kinds remained as they were originally created and there could therefore be no such things as homologous structures (p. 111). In constructing his *Scala Naturae* (ladder of life) Aristotle divided animals into *enaima* (= vertebrates) having red blood and reproducing by either vivipary or ovipary; and *anaima* (= invertebrates) with no red blood and viviparous, 'vermiparous' or spontaneously created. Criteria such as the colour of the blood and the presence or absence of wings were arbitrary and relied on purely superficial resemblances. In modern terminology, they were based on analogy rather than homology. It is worth adding that the words genus and species, which are employed universally today, are direct translations of the terms used for a similar taxonomic purpose by the early Greeks.

In a sense, early Christian beliefs served to reinforce the taxonomy of Aristotle and his contemporaries, since ideas based on a belief in the literal truth of divine creation as described in the first chapter of Genesis (p. 138), gave powerful support to the idea of the fixity of species. Small wonder that there was little change in outlook until the seventeenth century, by which time our knowledge of plant and animal anatomy had heralded a more critical approach.

Among the early taxonomists, the Englishman John Ray (1628–1705) combined the latest anatomical discoveries in plants and animals with the traditional classificatory system of Aristotle. To this he added the accumulated knowledge of geographical distribution. Ray was an ordained minister and accepted the biblical account of the Creation. None the less, he was also well aware of the basic structural patterns that characterised the different taxonomic groups, although he did not attribute these to an evolutionary process. He accepted the range of variation within species that he observed in wild populations, but attributed it entirely to environmental causes such as the effects of temperature differences or variations in soil fertility. His philosophy is summarised in the statement that 'plants which differ as species preserve their species for all time, the members of each species having all descended from seed on the same original plant'.

The death of Ray almost coincided with the birth of the Swede Carolus Linnaeus (1707–78) who, at an early age, developed a passion for classification. He classified not only plants and animals but also minerals and even the diseases of man. Linnaeus adopted a systematic approach to the use of taxa, beginning with the larger groupings (classes) and breaking them down to orders, genera, and species, his criteria being largely anatomical. Thus, he grouped plants primarily on the number of stamens in the flower: those with one were the class Monandra, two the class Diandra, three the class Triandra, and so forth. But he is remembered particularly for his contribution to taxonomy at the lowest levels and the establishment of a universal system of **binomial nomenclature**, whereby every organism is supplied with two names indicating the genus and species to which it belongs.

Much of the Linnean system of taxonomy has survived to the present day and the tenth edition (1758) of his monumental work *Systema Naturae*, still provides a source of reference for the international standardisation of species names. Like Ray, Linnaeus believed in the fixity of species. However, he accepted intraspecific variation which he attributed to hybridisation. Although they were unaware of the significance of homology as we know it today, both Ray and Linnaeus based their taxonomic systems on a study of basic structural resemblances between similar organisms. Theirs can, therefore, be regarded as the first steps towards a **natural classification**, in contrast to that of Aristotle and his contemporaries which relied largely on analogous resemblances. The life and work of Linnaeus in Sweden overlapped that of the two distinguished French biologists, George Buffon (1707–88) and Jean Baptiste Lamarck (1744–1829), and it is to them we must turn for a movement away from the static ideas of Linnaeus towards a more dynamic and evolutionary view of the Universe (see p. 144). In a new setting, the concept of homology in taxonomy was destined to take on a deeper significance and gradually to achieve the importance we assign to it today.

## What is a species?

If we are to study the modern system of taxonomy in the context of evolution, there is some justification for beginning at the level at which natural selection can best be seen to exert its effects. This is the species. Bearing in mind its universal acceptance as the smallest classificatory taxon, we might imagine that its definition would be clear-cut. But this is not so. Part of the problem arises from the different uses to which the term 'species' is put. Thus, as an international means of identifying a particular sort of plant or animal, the species needs to be exact and unambiguous. But to serve the needs of students of biology, its wider implications need to be taken into account, such as the range of variation that can be accommodated under a single name. To satisfy both practical and theoretical requirements, a number of criteria are therefore needed for the definition of species.

## Genetic criteria

The most objective definition of a species is a group of individuals with a common gene pool. This implies that interbreeding within the group produces fertile offspring. In general, hybrids (crosses between species) are infertile such as that between the primrose, *Primula vulgaris*, and the cowslip, *P. veris*, which produces the false oxslip that is incapable of forming seed (p. 96). For practical reasons, genetic criteria are of little value in the field or in professional institutions of taxonomy such as museums, so other more pragmatic approaches are necessary.

## Morphological criteria

Structural characteristics such as colour, size, and shape are those most frequently used for identification in the field and elsewhere. Provided the characters used for comparison are homologous, such an approach is justified. Problems arise, however, due to the differing degrees of variation, both genetic and environmental, that are characteristic of all species. If the taxon is to have any usefulness, some clear difference must exist between one species and the next. This implies that any definition must be sufficiently broad to encompass an acceptable range of variation.

## Ecological criteria

Individuals of the same species have similar habitat preferences. Among plants the term **indicator species** is sometimes used for those kinds characteristic of certain ecological environments such as heathland, grassland, or sand dunes. However, considerable variations can occur within a species, such as the polyploid forms of the marsh bedstraw, *Galium palustre* (p. 64), where chromosome number is related to the degree of moisture that the plants can tolerate. Again, among animals, different species exhibit particular behaviour patterns enabling them to occupy some niches within a community but not others.

## Geographical criteria

Species occupy a distinct geographical range and are usually separated from kindred species by a clear-cut discontinuity such as a change in ecological conditions or a physical barrier such as a mountain range or a stretch of water. Some species require rather special circumstances for their survival and their distribution is therefore restricted. A typical example is the little Scottish primrose, *Primula scotica*, which colonises short turf in cold, wind-swept conditions near the sea. In Britain its distribution is restricted to North Scotland and Orkney, and it occurs nowhere else. In northern Scandinavia its counterpart is *P. scandinavica*, which is similar to *P. scotica* but is regarded as a distinct species.

### Palaeontological criteria

As we saw in Chapter 2, the application of taxonomic principles to fossils presents particular problems. Species occupy a distinct and limited range of geological time. However, owing to the incompleteness of the fossil record it is virtually impossible to obtain the total history of a species or a precise idea of the range of its genetic and environmental variation. A second problem derives from the fact that although fossils can provide evidence of the time when they occurred and their worldwide distribution, they give only an approximate idea of the kinds of ecological conditions in which they lived. It follows that the degree of precision inherent in the palaeospecies is a good deal less than that of the biospecies.

### The polytypic species

As the eighteenth-century taxonomists such as Ray and Linnaeus fully realised, the concept of species as the smallest basic taxon necessitates the acceptance of a quota of inherent variation. Its extent fluctuates greatly from one species to another. In Victorian times it was fashionable to assign the term **type specimen** to one that represented the approximate species mean and therefore served as a point of reference for taxonomists. The practice was, however, of limited value for it tended to obscure the true range of variance. Botanists have traditionally preferred to cope with the variability of plant species by introducing subgroupings such as subspecies, varieties, subvarieties, and forms. In so-doing they earned the title of 'splitters'. Zoologists, on the other hand, have tended to accept the wide range of variation occurring in animals and confined their categorisation to species and subspecies. Taxonomists of this kind are sometimes referred to as 'lumpers'. As Mayr has stressed [6.6], by their very nature all species contain varying numbers of different forms, which may occur as graded series or be discontinuous. Particularly among animals, the species taxon is essentially **polytypic**.

The process of splitting or lumping together groups of similar organisms can be applied at any level in the taxonomic hierarchy from subspecies upwards. Its justification is that all taxa are, in varying degree, arbitrary units and therefore open to a range of differing interpretation.

## Taxonomy above the species level

### The taxonomic system

One of the greatest contributions of Linnaeus was that he standardised once and for all the use of the existing **binomial system** of naming each organism; its first name referring to the **genus** (prefixed with a capital letter and derived from the Latin *genus* = kind); and the second the **species** (written with a small letter, from the Latin *species* = appearance). Both are always written in italics, which is indicated by underlining. Linnaeus also carried his taxonomic system further to include larger taxa. Thus, a collection of similar genera constituted an **order**; a group of similar orders a **class**; and a group of similar classes a **phylum**. It should be noted that, today, botanists and microbiologists sometimes use the word **division** instead of phylum. The largest classificatory group is the **kingdom**. Thus, in descending order of size, the taxonomic hierarchy is kingdom, phylum, class, order, genus, and species. But as we saw earlier, there are numerous minor subdivisions of this scheme. The **family** is often inserted between the order and genus and is a particularly useful taxon for field biologists. In zoology this has the ending -**idae** and in botany -**aceae**. The **tribe** is occasionally added below the family. Moreover, all these groupings may be further divided into subgroups from the phylum downwards, the subspecies being the commonest.

An example of how the taxonomic system works in practice is given in Table 6.3, where the classifications of a rabbit and a daisy are compared. In this example it was useful to introduce one subgroup (**subphylum**). In the animal this enabled a distinction to be made between the chordates (animals with a notochord) and the vertebrates (animals with backbones). Similarly among the plants, the subphylum separates the ferns and seed plants from the club mosses and horsetails.

**Table 6.3** Taxonomy of a rabbit and a daisy

| *Taxon* | *Rabbit* | *Daisy* |
| --- | --- | --- |
| Kingdom | Animalia (animals) | Plantae (plants) |
| Phylum | Chordata (chordates) | Tracheophyta (vascular plants) |
| Subphylum | Vertebrata (vertebrates) | Pteropsida (ferns and seed plants) |
| Class | Mammalia (mammals) | Angiospermae (flowering plants) |
| Order | Lagomorpha (rabbits and hares) | Dicotyledonae (two cotyledons) |
| Genus | *Oryctolagus* | *Bellis* |
| Species | *cuniculus* | *perennis* |

### Significance of the larger taxa

Calow [6.7] has stressed the important point that when considering the higher levels of the classificatory hierarchy in the light of their evolutionary significance, two aspects need to be taken into account:

(i) the order in which descendants have arisen from a common ancestry, and
(ii) the extent to which divergence has taken place during the process of descent.

Suppose that several related species have an affinity with one another and their ancestor, and have diverged relatively little during their evolution. Clearly, there would be a strong argument for classifying them together. But if one of those species should exhibit a major deviation from the ancestral pattern, should it still be grouped with the other descendants or classified separately? In other words, in applying our taxonomic principles, to what extent should we think vertically and horizontally, and which, if either, pattern should take precedence?

### Cladistics

One school of taxonomy subscribes to a scheme originally formulated by the German biologist W. Hennig in 1950, based on the idea of **genealogy** and known as **cladistics** (derived from the Greek word *klados* = branch or shoot). The basic taxon is the **clade** and organisms grouped within it are assumed to have been derived from the same ancestor. An important assumption in the construction of cladistic diagrams (**cladograms**) is that all species present at a particular time are not themselves ancestors of other existing species. It follows, therefore, that they must be related, if somewhat distantly by descent from other ancestral forms.

Cladograms are constructed on the basis of a number of shared characteristics which are considered homologous and therefore indicate an evolutionary relationship. Moreover, the similarities must exist at the assumed branching point of each clade. The principle has been well illustrated by Patterson [6.8]. Suppose we wish to classify two different kinds of chickens (A, B) and a goose (C). An appropriate cladogram is shown in Fig. 6.8. The two chickens are linked together because they show homologous characteristics that are absent in the goose. The chickens therefore form a clade AB to which the goose does not belong. On the other hand, the chickens and the goose share homologies extending to a wider group ABC, of which AB and C are subgroups.

Extending this principle a step further, suppose we wish to construct a cladogram to show the relationship of a man, cat, dog, and fox. Man, being a primate, shares certain characters with the other three such as the possession of pentadactyl limbs, a mammalian reproductive system and so forth, but differs in such features as bipedalism and a large brain. Here, then, is the first point of division:

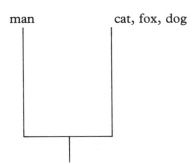

The fox and dog have numerous characters in common with one another and with the cat, but differ from it in such features as dentition. This provides the second point of division, and the final cladogram is:

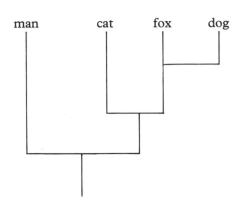

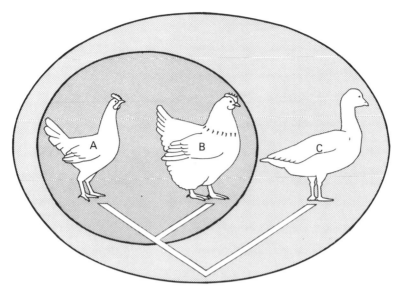

**Fig. 6.8** A cladogram of relationships between two kinds of chicken and a goose (after Patterson 1982)

When constructing a cladogram, one of the greatest problems is to determine whether the characters used for comparison are genuinely homologous as between the different species (sometimes known as **derived characters**). If similarities within a cladist group are also shared with organisms outside it, such as pentadactyl limbs in the previous example, these are probably not derived and are referred to as **primitive characters.**

## Phenetics

A different school of thought rejects the idea that classification can be based on genealogy. This is because it is often difficult in practice to establish genuine homology and to distinguish it from the results of convergent evolution and structures exhibiting analogy only, such as the limbs and wings of insects, and vertebrates. Phenetics involves the use of a wide range of characteristics covering overall similarities and differences without reference to homology and analogy; the argument is that if enough characters are used, the effects of homology should outweigh those of analogy. If this is so, a phenogram (Fig. 6.9) should give an indication of evolutionary relationships.

An alternative viewpoint is that classification above the species level is no more than a convenient means of pigeonholing organisms that does not necessarily reflect patterns of evolution. Biologists such as Sneath and Sokal have worked out complex systems for assigning numerical values to different degrees of resemblance – known as **numerical taxonomy** [6.9]. A simple illustration of the principle involved in constructing a phenogram is shown in Fig. 6.9. Suppose we wish to classify five organisms, ABCDE; the scale of arbitrary units in ascending order is on the right of the diagram. A and B closely resemble one another and are assigned a value of 8; C and D are also similar but less so, and score 7. They are therefore represented on the phenogram as shown in (a). The resemblance

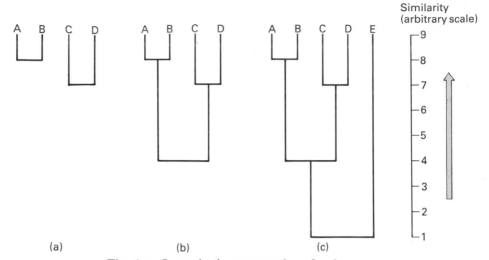

**Fig. 6.9** Stages in the construction of a phenogram

between AB and CD is more divergent and amounts to five units, as is illustrated in stage (b). Organism E differs considerably in overall characteristics from ABCD with a similarity score of 1 as is shown in (c).

## Phylogenetic trees

Tree-like branching diagrams are often regarded as the traditional means of expressing relationships, and involve the use of supposed genealogy and overall similarities and differences in expressing divergence. The procedures for determining phylogenetic relationships have been elaborated by Mayr [6.10]. They are less objective than those of cladistics and phenetics, and depend to a considerable extent on the breadth of knowledge and judgement of those concerned. A good example has been quoted by Calow [6.7] and concerns the taxonomy of the reptiles and birds (Fig. 6.10). The latter were undoubtedly derived from an ancestral

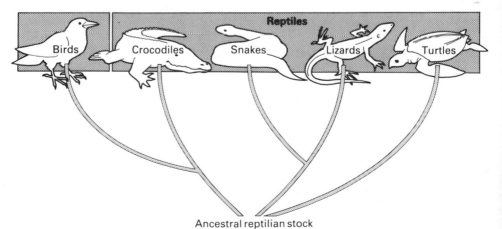

**Fig. 6.10** Phylogenetic tree showing the likely ancestry of the reptiles and birds (after Calow 1983)

group of small dinosaurs that flourished during the Jurassic and Cretaceous periods. The lineage of birds is closer to crocodiles than to other reptiles; hence they have more characters in common than the crocodile shares with other living reptiles. Once they had separated from the reptilian line, the birds diverged rapidly and the Jurassic fossil *Archaeopteryx* possessed feathers and well-developed wings. Birds have diverged further from the reptilian stock than have crocodiles and are therefore grouped in a separate class Aves. The crocodiles, however, are grouped with the other reptiles in the class Reptilia.

Phenetics would arrive at a similar answer on the basis of the quantitative assessment of structural similarities. However, cladists would base classification on genealogical resemblances only and therefore construct a cladogram showing the birds and crocodiles to be closely related and the snakes, lizards, and turtles separated from them.

## An ideal taxonomic system?

From the previous account it will be clear that establishing evolutionary relationships between organisms is no easy matter. Homology among structures and processes is often difficult to detect and highly speculative. Convergent evolution between distantly related groups has frequently occurred in the past leading to varying degrees of analogy. Moreover, there seems to be no *a priori* justification for the assumption made by some pheneticists that the larger the number of characters included, the greater the likelihood that the influences of convergence will be swamped by those of homology.

So the conclusion must be that in our present state of knowledge, there is no ideal scheme for determining the taxonomy of plants and animals. Opinions will vary as to the nature and composition of different groups. One procedure would be to establish a new taxon at each level of branching on the evolutionary diagram, but this, as we have seen, can lead to differing results. Ultimately, the different assemblages have somehow to be fitted into the Linnean scheme of phyla, classes, orders, and genera. Bearing in mind the degree of subjectivity inherent in the process, it is not surprising to find, particularly in the lower levels, that the taxonomic system is subject to frequent controversy and adjustment.

Perhaps the most important outcome of a discussion of current taxonomic principles is that it serves to highlight the need to relate the process of macroevolution to that of microevolution. To what extent are situations described in the previous pages explicable in terms of neo-Darwinism; do we need to invoke the influence of different processes to account for them? These are fundamental questions to which we will return when considering the modern synthesis in Chapter 8.

## The ancestry of man

In the previous pages we have purposely avoided a discussion of the taxonomy of any particular group of plants or animals, since our main concern has been with taxonomic principles and the problems they pose in the study of evolution above the species level. In conclusion, it will be worth treating the ancestry of man as a separate issue, partly because of its own intrinsic interest, and partly because it illustrates well the application of the taxonomic methods enunciated earlier. It also introduces an increasingly important additional line of evidence on homology derived from comparative cytology and biochemistry.

### Origins of primates

Human beings (genus *Homo*) are classified in an order of mammals known as Primates which are thought to have made their appearance in the late Cretaceous. This was a period of rapid evolution among the class Mammalia, involving adaptive radiation (p. 108) in numerous different directions and adjustment to a wide variety of ecological niches. The approximate relationship of some of these forms has been summarised by Campbell [6.2] and is shown in Fig. 6.11. The early primates evidently lived in forests and were tree-dwellers, so that two particular attributes will have been at a selective advantage – limbs with a high mobility, and effective vision. Some typical primate characteristics are:

  (i) Increased flexibility of the generalised pentadactyl limb.
 (ii) Flexible digits with nails to facilitate grasping.
(iii) Retention of the tail as an organ of balance.
 (iv) An upright (bipedal) posture with the ability to rotate the head.
  (v) Elaboration of the nervous system, particularly the size and complexity of the brain.
 (vi) Enlargement of eyes and development of binocular vision with overlapping visual fields and the ability to judge distance.
(vii) Development of the frontal region of the bony skull as a protection for the eyes.
(viii) Correlated with the increase in vision, a reduction in the sense of smell associated with smaller olfactory lobes in the brain and a shortening of the snout.

Some typical living primates, both primitive and advanced, are included in Fig. 6.12.

We are not concerned here with a detailed taxonomy of the primates but only with the ancestors of man – the monkeys and their allies. These are divided into the New World monkeys (**platyrrhines**) found in Central and South America,

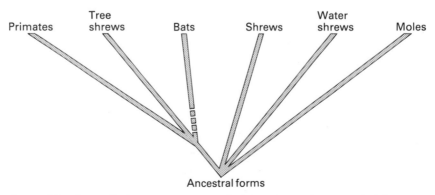

**Fig. 6.11**  Adaptive radiation among some primitive placental mammals (after Campbell 1976)

and the Old World monkeys (**catarrhines**) distributed in Africa and Asia. Anatomical and other evidence shows clearly that human beings are related to the catarrhines. Their elaborate structure and basic similarities place them with the monkeys and apes as 'higher' primates, in the suborder Anthropoidea. Within this group is the superfamily Hominoidea to which we belong, together with the apes.

In Table 6.4 is set out in the left-hand column a sample of the range of characteristics thought to be homologous and now used to establish taxonomic links among anthropoids, particularly with man's closest living relatives – the gorilla and chimpanzee (Fig. 6.13). To these must be added the remarkable findings, obtained as long ago as 1939 [6.11], from testing anthropoid apes for their ability to taste the bitter compound phenylthiocarbamide – a characteristic inherited as a dominant in man. In the sample of chimpanzees tested, not only were the tasters and non-tasters readily detectable, but the characteristic also occurred in approximately the same proportions (roughly 26 per cent non-tasters) as in human populations. As the researchers point out, it seems almost inconceivable that over the million or so generations since the separation of the anthropoid and hominoid stocks, the gene ratio should have remained unchanged. That it has evidently done so indicates that during this vast period ability to taste this odd chemical substance has continued to enjoy a selective advantage. The precise reason for the perpetuation of this peculiar dimorphism still remains

**Fig. 6.12**  Some living primates: (a) tree shrew, (b) lemur, (c) tarsier, (d) Old World monkey, (e) chimpanzee, (f) Australian aboriginal

**Table 6.4** Homologous resemblances and differences in the gorilla, chimpanzee, and man

| Characteristics | Gorilla | Chimpanzee | Man | Cladogram |
|---|---|---|---|---|
| *Bones and teeth* | | | | |
| Limb length | Legs shorter than arms | Legs shorter than arms | Legs longer than arms | (c) |
| Canine teeth | Large | Large | Small | (c) |
| Thumb length | Short | Short | Long | (c) |
| *Chromosomes* | | | | |
| Diploid number (2N) | 48 | 48 | 46 | (c) |
| Nos 5 and 12 | Different from other primates | Different from other primates | Like other primates | (c) |
| No. 13 and Y (fluorescence in ultraviolet light) | Same as man | Like other primates | Same as gorilla | (a) |
| Inversions | Differ from man by eight inversions | Differ from man by six inversions | — | (b) |
| Banding | Similar to man | Differs from man and gorilla | Similar to gorilla | (a) |
| *Biochemistry* | | | | |
| Blood plasma proteins (fibrinopeptides A and B) | Same as man | Same as man | — | (a) or (b) or (c) |
| $\alpha$-and $\beta$-haemoglobin | Differs by one amino acid | Same as man | Differs by one amino acid | (b) |
| $\delta$-haemoglobin | Same as chimpanzee | Same as gorilla | Differs by one amino acid | (c) |
| Myoglobin | Same as chimpanzee | Same as gorilla | Similar to chimpanzee | (c) |
| DNA structure | Differs from man and chimpanzee | Similar to man | Similar to chimpanzee | (b) |

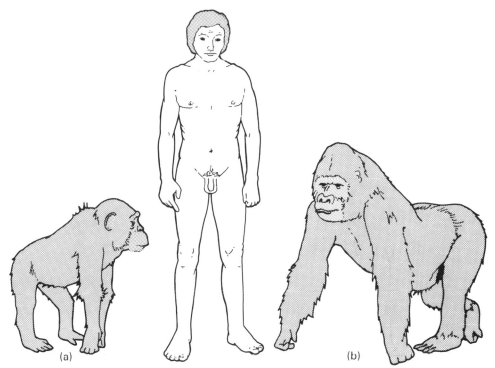

**Fig. 6.13** Man and his nearest living relatives: (a) chimpanzee, (b) gorilla

obscure, but there is statistical evidence of an association with two diseases of the thyroid gland – Graves' disease, which is commoner among tasters, and adenomatous goitre, which is more frequent in non-tasters.

It will be seen that the criteria used for comparison in Table 6.4 include not only structural resemblances such as bones and teeth, but cytological characteristics relating to chromosome number and composition, and biochemical features such as blood components and the structure of DNA. Moreover, the degree of relationship between the three anthropoids varies depending on which characteristics are used to separate them. Some indicate that man and gorilla are more closely related to one another than to the chimpanzee; others that man and chimpanzee are closer to each other than to the gorilla; and again that chimpanzee and gorilla are nearer together than to man. These alternative possibilities are presented as three cladograms in Fig. 6.14 and related to the different characteristics in the right-hand column of Table 6.4. Summing up the results, two items

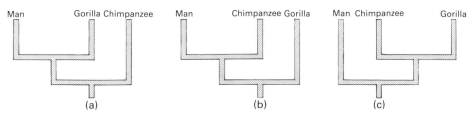

**Fig. 6.14** Cladograms showing alternative relationships of the gorilla, chimpanzee, and man

favour cladogram (a); three favour (b); and seven suggest (c) – with one uncertain. Clearly, not all the criteria are of equal value; some are more relevant than others. The difficulty is to be sure of their relative significance; and in our present state of knowledge this is not possible. The conclusion must therefore be that although there is strong evidence that chimpanzee and gorilla have a closer affinity to one another than to man, it is wiser to regard both as our closest living relatives and look elsewhere for links between them and modern man.

### Fossil species of man

Following excavations in localities as far apart as Western Europe, Africa, China, and Indonesia, we now have a much clearer picture of the close relatives of modern man. For convenience, we will consider only those species included in the genus *Homo*, to which human beings (*Homo sapiens*) belong.

    *Homo habilis* inhabited East Africa 1.5–2 million years ago (Fig. 6.15) and

**Fig. 6.15** Reconstruction of *Homo habilis*

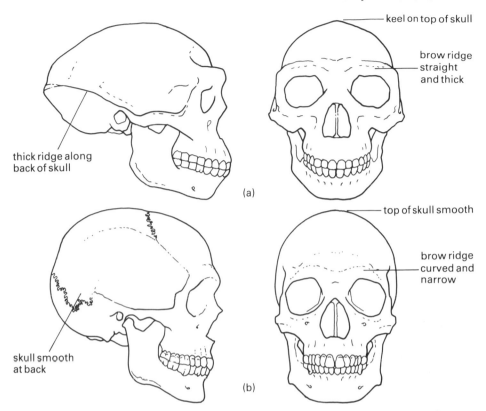

keel on top of skull

brow ridge straight and thick

thick ridge along back of skull

(a)

top of skull smooth

brow ridge curved and narrow

skull smooth at back

(b)

**Fig. 6.16** Reconstruction of the skull of *Homo erectus* (a) compared with that of modern man, *Homo sapiens* (b)

possessed many features that are typically human, such as bipedalism and a flattened face with only a short snout. From the size of their bones, it has been estimated that the average weight of habilines must have been around 40 kg (85 lb). From the shape of the skull it is clear that, for their size, the brain was large and well developed. Evidence of this is provided by the fact that *H. habilis* is the first known maker of tools. These consisted of implements made by knocking one stone against another to form a cutting edge.

Another early human species, *Homo erectus*, had a wide distribution, occurring first in East and South Africa and later spreading to northern Asia, the East Indies, and Europe. Like the habilines, individuals had a typically flattened human face, were bipedal, had a large brain and fashioned stone tools. Relative to their bodily size, the brain was further developed than in *Homo habilis*, and their mode of life included one major advance – the use of fire. However, a comparison of the skulls of *H. erectus* and modern man, *H. sapiens* (Fig. 6.16), shows some significant differences. Note particularly in the *H. erectus* side view the ridge along the back of the skull; and from the front, the keel at the top of the skull for the attachment of muscles, and the thick ape-like brow ridges and high cheek bones.

Thus, although both species of fossil man have close affinities with ourselves, the characteristic differences suggest that neither should be regarded as our immediate ancestor. If we include the apes, discussed earlier, the appropriate cladogram is probably as in Fig. 6.17.

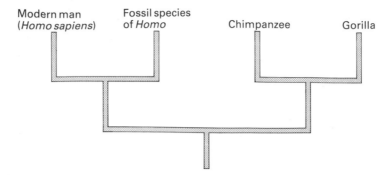

**Fig. 6.17**　Cladogram showing the probable relationship of the gorilla and chimpanzee with modern man and fossil species of *Homo*

Our nearest known relatives colonised much of Europe and the Middle East between 100 000 and 40 000 years ago. In many ways they resembled modern man, particularly in having an average brain capacity of about 1330 cm³. They carved implements beautifully, used fire and buried their dead with due ceremony. On such grounds, they are grouped in the same taxon as ourselves – *Homo sapiens*. However, there are also clear-cut differences, as a comparison of the respective skulls shows (Fig. 6.18). In particular, the prominent brow ridges that are a feature of other fossil species still persist, but in reduced degree, and the sloping forehead behind them. For such reasons, these early human beings are referred to as a subspecies, *Homo sapiens neanderthalensis*, named after the valley of the River Neander in Germany where remains of them were first discovered in a remote cave. The earliest known human inhabitants of Britain were similar to Neanderthals. Their remains have been found at Swanscombe in Kent and date to about 250 000 years ago.

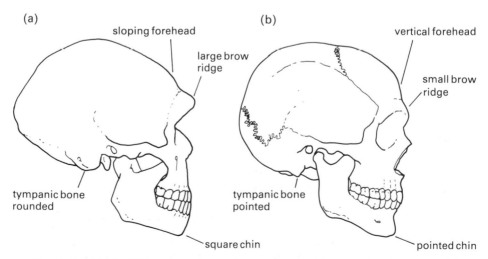

**Fig. 6.18**　Skull of Neanderthal man (a) compared with that of modern man (b)

## The problem of human races

The word 'race' is often used in a loose sense to describe variants of *Homo sapiens* by their appearance. Thus, we speak of Caucasoids with fair skins such as Europeans, Mongoloids with light-brown skins and dark hair such as the Chinese, the Negroids of Africa, and the Australoid aboriginals. Such human varients differ less from one another than did Neanderthals, and are therefore not accorded subspecies status. From a taxonomic viewpoint, we can define a race as a separate group of a species with a distinct geographical distribution and distinguishable from other races. The differences between races sometimes reflect adaptations to particular environmental conditions. The indigenous populations of tropical countries such as Africa, for instance, are predominantly black skinned, the pigment melanin providing protection from an excess of ultraviolet light that could otherwise cause a high incidence of skin cancer. For other characteristics, the action of natural selection over time has been more subtle and we are still far from clear about their significance. For instance, the human blood groups, particularly the ABO series, undoubtedly denote racial differences which have already been discussed in the context of polymorphism (p. 116). Some, at least, have now been shown to confer certain advantages and disadvantages in relation to medical conditions such as endemic disease.

Patterson [6.12] has drawn attention to the interesting fact that different kinds of data can lead the user to widely divergent conclusions when attempting to trace the relationships of human races. This is illustrated in Fig. 6.19. Cladogram (a) is

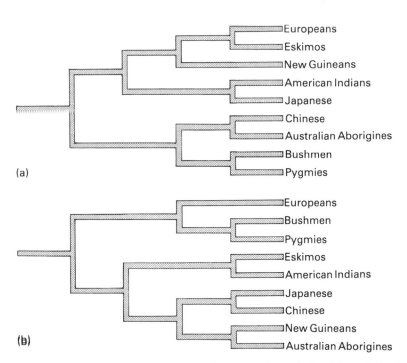

**Fig. 6.19** Cladograms showing the supposed genealogical relationships of nine human races: (a) based on twenty-six anatomical features, (b) based on fifty-eight genetic markers (from Patterson 1978)

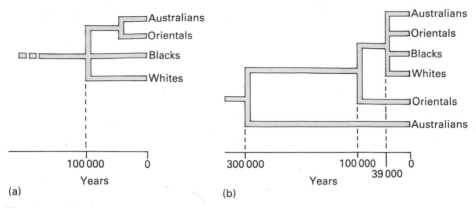

**Fig. 6.20** Cladograms showing the possible relationships and time of origin of four human races: (a) based on blood proteins and nuclear DNA, (b) based on mitochondrial DNA (from Gribbin and Cherfas 1982)

based on such features as eye, hair, and skin colour, limb proportions and facial characteristics. Cladogram (b) is derived mainly from biochemical variants such as differences in blood proteins. Which of the two ancestral trees is likely to be nearer the truth? The characters used in (a) are those most likely to be influenced by environmental considerations such as temperature, illumination, and the requirements for different kinds of activity. Those in (b) are more fundamental to the species; although they may have been subject to natural selection, they are not likely to have been modified to the same degree and they therefore provide the more reliable evidence of affinity.

But biochemical evidence of racial differences and origins can also be conflicting, as Gribbin and Cherfas [6.13] have described. In Fig. 6.20 are two cladograms showing the supposed genealogy of four human races set against a time scale. One (a) is derived from data provided by blood proteins and nuclear DNA; the other (b) is based on mitochondrial DNA. It will be seen that there is substantial agreement between them but the mitochondrial evidence in (b) shows that offshoots of the Australians and Orientals occurred a good deal earlier than is suggested by their nuclear DNA and blood proteins. Whatever model we care to choose to explain such discrepancies, the unfortunate fact remains that the concept of race is invariably tinged with political overtones. From the previous account one clear conclusion emerges, that no matter what views we eventually arrive at, they are bound to remain only speculative and therefore open to doubt.

## Summary

1 Microevolution covers the study of evolution at species level and below. Macroevolution is concerned with evolution above the species level.

2 The fossil record, adaptive radiation, the effects of isolation, and domestication provide evidence for evolution.

3 Homology occurs when characteristics present in two or more species can be traced to a common origin. It provides the basis for a natural type of classification.

4  Embryology provides evidence of evolution through recapitulation, the existence of vestigial structures, and alterations in the relative growth rates of different structures. These can lead to conditions such as neoteny (precocious sexual maturity).

5  The geographical distributuion of plants and animals reveals evidence of patterns of evolution resulting from isolation and the development of endemic species.

6  Physiological similarities such as the blood groups, blood proteins, and the composition of DNA can provide valuable evidence of affinity and differences between living organisms.

7  The modern scheme of taxonomy owes much to Ray and Linnaeus who established the binomial system of genera and species, also the higher taxa. Classification follows a natural process in that it relies greatly on homology.

8  The species constitutes the smallest classificatory taxon. The members of a species are determined by genetic, morphological, ecological, geographical, palaeontological and other criteria.

9  The variation occurring within species leads to the concept of polytypic groupings of organisms.

10  The larger taxa are arbitrary in their composition and rely on the identification of homologous features. Three different approaches are used in establishing taxonomic relationships – cladistics, phenetics, and phylogenetic trees.

11  Cladistics seeks to establish genealogy through the identification of shared characteristics and the construction of cladograms. The main problem is to distinguish genuine homology from apparent relationships due to convergent evolution.

12  Phenetics depends upon the use of numerous detailed characters covering overall differences without reference to homology or analogy. The assumption is that, on average, the effects of homology will outweigh those of analogy. However, there is no certainty that this assumption is always justified.

13  Phylogenetic trees are more generalised and concerned with overall similarities thought to be related to evolutionary divergence, the majority being homologous.

14  The ancestry of man provides a good example of the application of different evolutionary criteria and a comparison of findings based on biochemical and anatomical evidence.

15  Biochemical characteristics have been particularly useful in establishing the relationship of modern man with apes such as the gorilla and chimpanzee. They have also provided evidence on the evolutionary affinity of different human races.

## Topics for discussion

1  It is generally assumed that human evolution occurred on land, but an alternative hypothesis suggests that we have a marine ancestry. What kind of evidence would you look for to support or disprove the idea that man evolved from the sea?

2  What do we mean when we say that man's evolutionary success has been made possible by our lack of specialisation?

3 Man is a neotenous ape. What evidence is there that this is so?

4 To what extent does a hierarchical scheme such as a cladogram, phenogram or phylogenetic tree provide evidence for organic evolution?

5 When tracing the evolutionary relationships of a group of organisms, it was found that anatomical evidence pointed to one conclusion and biochemical evidence to another. Which is likely to be the more reliable and why?

6 During its development, an animal ascends its ancestral tree. To what extent can this statement be regarded as true?

7 What evidence for man's evolution is provided by our wisdom teeth?

8 What criteria can we use in deciding whether resemblances between organisms are analogous or homologous?

9 What deductions can be drawn from the fact that the British Isles have very few endemic species of plants and animals while islands such as Hawaii have a great many?

10 The existing taxonomic system suffers from wide variations in interpretation. What would be the essential features of an 'ideal' taxonomy and why is this unlikely to be achieved?

# References

**6.1** de Beer, G. R. (1954) *Embryos and Ancestors*. Oxford University Press

**6.2** Campbell, B. G. (1976) *Human Evolution*, 2nd edn. Heinemann

**6.3** Tivy, J. (1982) *Biogeography*, 2nd edn. Longmans

**6.4** Ford, E. B. (1973) *Genetics for Medical Students*, 7th edn. Chapman and Hall

**6.5** Clarke, C. A. (1982) 'The uncertainty principle as applied to medicine', *Journal of Biological Education* **16**, 93–6

**6.6** Mayr, E. (1964) *Systematics and the Origin of Species*. Dover, New York

**6.7** Calow, P. (1983) *Evolutionary Principles*. Blackie

**6.8** Patterson, C. (1982) *Cladistics*. In *Evolution Now: a Century after Darwin* (ed. J. Maynard Smith). Nature/Macmillan

**6.9** Sneath, P. H. A. and Sokal, R. R. (1973) *Numerical Taxonomy*. W. H. Freeman, San Francisco

**6.10** Mayr, E. (1970) *Populations, Species and Evolution*. Harvard University Press

**6.11** Fisher, R. A., Ford, E. B., and Huxley, J. (1939) 'Taste-testing in the anthropoid apes'. *Nature* **144**, 750

**6.12** Patterson, C. (1978) *Evolution*. Routledge and Kegan Paul

**6.13** Gribbin, J. and Cherfas, J. (1982) *The Monkey Puzzle*. Bodley Head

# 7

# Alternative explanations of change

The debate generated by the proposals put forward by Darwin in *The Origin of Species* was intense and sometimes bitter, and soon spread beyond Britain to the United States, Russia, and countries throughout Europe. The claim that natural selection, acting upon the fund of variation that existed in all wild populations, could be responsible for the evolution of new species upset many preconceived ideas, and it is not surprising that criticisms and objections were soon forthcoming from many quarters. In general, the criticisms were of two kinds – religious on the grounds that Darwinism was incompatible with established church doctrine and scientific because the new approach deviated fundamentally from existing views. It must also be remembered that many of the leading scientists of the day, some of whom were opponents of Darwin, held strong religious convictions, so that his ideas were objectionable to them on both counts.

## Darwinism and religion

Some of the most vicious critics of Darwin represented a combination of science and the Church, such as the biologist Richard Owen and Samuel Wilberforce. Bishop of Oxford (widely known as Soapy Sam on account of his patronising manner). Powerful advocates of Darwin included the distinguished biologist T. H. Huxley. Huxley possessed all the attributes that Darwin lacked. Thus, while Darwin was of a retiring disposition and a poor public speaker, Huxley had the reputation of being an outstanding university lecturer and brilliant debater. His outspoken advocacy of the views expressed in the *Origin*, earned him the name of 'Darwin's bulldog'.

The epic clash between these two opposing forces occurred at the meeting of the British Association for the Advancement of Science at Oxford on the 30 June 1860. Wilberforce (with the backing of Owen) read a paper aimed at Huxley and the Darwinist viewpoint on evolution and in concluding was unwise enough to pose the question as to whether it was through his grandfather or his grandmother that he (Huxley) claimed descent from a monkey. This prompted the now famous retort:

> If then, said I, the question is put to me – would I rather have a miserable ape for a grandfather or a man highly endowed by nature and possessed of great means and influence, and yet who employs these faculties and that influence for the purpose of introducing ridicule into a grave scientific discussion – I unhesitatingly affirm my preference for the ape.

Huxley had won the day but not everyone was happy with the outcome. Thus, it is said that when the wife of the Bishop of Worcester was told by her husband what had happened she exclaimed, 'Descended from the apes! My dear, let us hope that it is not true, but if it is, let us pray that it will not become generally known' [7.1].

These examples of the antagonism generated by *The Origin of Species* serve to highlight the issues on which evolutionists and churchmen were in conflict. In varying degree, that conflict has existed ever since. Nowhere has this been more intense than in certain parts of the United States, where a movement sprang up soon after the First World War among a number of different religious denominations known as **fundamentalism**. Originally, a **fundamentalist** was one who adhered to the fourteen fundamental doctrines laid down in the Niagara Bible Conference of 1878. These included among other requirements, a belief in the literal truth of the Creation as described in the Book of Genesis. More recently, the use of the term fundamentalist has become increasingly widespread and its original meaning blurred. Other aspects of the 1878 creed have been omitted and only the belief that Genesis represents a true account of Creation remains. It is in this sense that the term fundamentalist will be used in the account that follows.

Darwin and his associates were now suggesting that new species could arise and had arisen in the past as a result of an observable scientific process. Particularly objectionable to the Victorians was the idea that the origin of man was part of this process. Although few references are made to this possibility in the *Origin*, Darwin went out of his way to develop the theme in a subsequent book, *The Descent of Man* published 12 years later, in 1871.

As de Beer [7.2] has pointed out, Darwin fully appreciated the implications of the evolution of the human brain, which were first drawn to his attention by Wallace in 1864. The elaboration of man's mind served to explain why the human body is devoid of weapons such as teeth and claws for offence and defence. The brain had superseded the body as the principal mechanism for survival through the evolution of intelligence, enabling prey to be captured and enemies outwitted. Moreover, if weapons were needed, man was now capable of devising them.

What Darwin failed to appreciate fully was the significance of the evolution of speech and the capacity for communication which, aided by higher mental faculties, had resulted in the accumulation of knowledge through experience. This could be passed on from one generation to the next. The process is quite distinct from Darwinian evolution, which depends upon the differential survival of phenotypes and the transmission of their genes to their descendants through the process of reproduction. The evolution of man had thus brought with it a new kind of change, based not on differing ecological conditions but on the varied characteristics of society, including its religious convictions. Hence, it came to be known as **psychosocial evolution**. Attempts to study its occurrence and outcomes cannot be divorced from political considerations, for the nature of societies and differences in their ethical standards depend to a considerable extent on, for instance, whether they are democratic or totalitarian.

What, then, should be the scope of the word 'evolution'? Should it be confined to organic (biological) evolution or should it embrace psychosocial evolution as well? There is no easy answer to this question. However, although few scientists doubt that organic evolution is the province of biologists, some would restrict psychosocial evolution to anthropology or sociology.

## Evolution and Christianity today

It is not easy to summarise in a few words the differing views of Christians on evolution and its implications, particularly those for man. But the considerations that follow would seem to be essential items for the agenda or any debate on the present relationship between science and the church.

Unlike the time of Darwin, fundamentalism today is not the strong force in Britain that it used to be, and relatively few Christians subscribe to the literal truth of the account of the Creation as recorded in the Book of Genesis. Those who do, usually reject the idea of organic evolution. The situation is of particular interest in certain areas of the United States where strongly held fundamentalist beliefs have resulted in significant political, social, and educational repercussions following the development of so-called **creationist science**. We will return to this issue later in the chapter (p. 141).

Most Christians, if they think about evolution at all, would be prepared to accept the findings of science in the field of science. For them the words of the agnostic T. H. Huxley, 'follow humbly wherever nature leads', and those of the clergyman Charles Kingsley, 'science is the voice of God', point in much the same direction. Many of the statements about biology contained in the first three chapters of Genesis are incorrect. But this need not detract from the account as an allegorical act of praise, which is what its writers probably intended it to be. There are comparable poetic descriptions of the origins of life in several of the other great religions of the world, such as Islam and Buddhism. Again, regarding the Creation, some Christians would point to the uncertainty and speculation that still enshroud the origin of life, as we saw in Chapter 1. And even if these were to be resolved there is still the far greater mystery of the origin of the Universe.

When referring to the 'Christian' viewpoint a warning note is necessary, for the doctrines of the different churches concerning the same issues can vary considerably. As was suggested earlier, most members of the Church of England do not hold strongly fundamentalist views. But for many Roman Catholics, the doctrine of original sin refers directly to a sin committed, in actual historical fact, by an individual man named Adam. Its relevance today therefore derives from the assumption that we are all descended from him.

Related to creationism is the concept of purpose and the idea of a divine plan for the development of the Earth and mankind upon it. As we saw in Chapter 2, the **determinist** approach has little evidence to support it, for while data from fossils show that certain groups such as the horses evidently evolved in a more or less linear manner (sometimes referred to as **orthogenesis**), there is nothing to suggest that this was part of a wider scheme nor that there was any ulterior purpose or goal. The theory of evolution derives from scientific facts and deductions made from them, and is therefore essentially **materialistic** in outlook.

But there is another side to our make-up, the moral and ethical one, that poses a different set of evolutionary problems. Can the development of attributes that, as far as we know, are uniquely human, such as self-awareness, the ability to reason, and to distinguish between good and evil, be accounted for purely in terms of their survival value? Have they an adaptive significance in a Darwinian sense? At this point materialism has little to offer and we move into a field of values separate from the material world. To the Christian, ethical and moral values such as those enunciated in the Ten Commandments and elsewhere, and personified in the life of Jesus Christ, are regarded as God-given. They have been revealed to mankind in the words of the Bible and thus comprise an essential component of his faith. To the agnostic, such values are the outcomes of a society that has evolved over time and are therefore products of the process of psychosocial evolution. The pursuit of these ideas further leads us into the complex field of evolutionary mysticism where we are faced with yet another set of ideas that must be classed as nearer to philosophy than science [7.3].

The existence of the bodily and spiritual facets of man has prompted the question whether we should, in reality, be regarded as two beings: a body subject to natural selection and the laws of organic evolution and a 'soul', which is the product of divine creation. If this were so, the question would arise whether the expression of patterns that are 'divine' can be modified both in degree and kind, as they invariably are in different human beings.

## Altruism and man

At different levels of the animal kingdom we find a kind of behaviour which is essentially unselfish and may even involve self-destruction for the benefit of others of the same species. Such behaviour is known as **altruism** and poses interesting problems of evolution. Put more precisely, an altruistic animal is one that reduces its own fitness to survive in order to increase the fitness of another. Among lower animals such as insects, altruistic behaviour is instinctive; there is no question of choice. But in higher forms such as primates a readiness to sacrifice oneself in the interest of others requires a decision. Monkeys such as baboons live in close-knit social groups or troops. In the presence of a potential predator such as a leopard, the dominant male will issue a defensive challenge on behalf of the troop that could cost him his life. The troop, meanwhile, escapes unharmed.

It has been postulated by Dawkins [7.4] and others as an axiom of population genetics that genes promoting the capability of an animal to behave altruistically will be at a selective advantage provided the loss they cause to the individual carrying them is offset by the benefit received by the relatives. If this is so, the more distant the relative the greater will be the improvement required. Brothers and sisters have, on average, roughly half their genes in common. For a gene to be favoured by selection that resulted in one brother sacrificing his life for another, the surviving brother would need to raise more than twice as many children as he would otherwise have done before the altruistic act. First cousins share, on average, about a quarter of their gene complement. So for altruism by a cousin to have positive value, a fourfold increase in fitness would be necessary. Within a community, altruism can occur simultaneously at different levels, by parents in favour of their young, among the young themselves, and so forth. The resulting benefits then represent the sum of the altruistic acts quantified according to their genetic significance. This process is sometimes known as **kinship selection**. There seems little doubt that it has played a part in the evolution of the more elaborate forms of social behaviour, such as those occurring among the primates and, particularly, man.

But in human societies altruism has been taken a stage further. Not only are men and women prepared to sacrifice themselves for their relatives, but if necessary they are also ready to die for causes bearing little or no direct relationship with the family. Such behaviour differs from that of lower animals in that it is not instinctive but derives from conscious decisions. Wars, rebellions, and riots have been and still are all too familiar features of man's existence. Their causes are usually social, economic, or religious, all of which clearly fall into the psychosocial domain that we considered earlier (p. 139). In terms of selective advantage, patriotism in wartime is a mixed blessing. Although the outcome may result in benefits for the population concerned, the cost in deaths could also be regarded as advantageous in imposing a temporary interruption to an unwelcome population increase. On the other hand, a high mortality among its young men will deprive a nation of an important component of its gene pool that it can ill afford to lose.

## Creation science

In the previous discussion of the influence of fundamentalist ideas on our concept of evolution, the point was made that at the time of Darwin there was a common belief in the literal truth of the Bible. Since then, fundamentalism in Britain has declined to the extent that it is no longer a significant influence on evolutionary thought. In the United States, the course of events has been different. A fundamentalist movement started to gather momentum towards the end of last century and reached a watershed in the now famous Tennessee 'monkey trial' of 1925. In this, a school teacher named Scopes was held to have infringed a state law known as the Butler Act which made it unlawful to teach 'any theory that denies the story of the divine creation of man as taught in the Bible, and to teach instead that man was descended from a lower form of animals'. Scopes was at first convicted but appealed to the Supreme Court on a technicality and his conviction was eventually quashed. The trial generated enormous public interest comparable with the Huxley–Wilberforce debate some 85 years earlier [7.5]. The outcome of the 'monkey trial' was thought at the time to signal the end of the anti-evolution lobby. But in 1981, we find a law being introduced in Arkansas (the Balanced Treatment of Creation Science and Evolution Science Act) requiring that 'Public (i.e. Maintained) Schools within this State shall give balanced treatment to creation science and evolution science'. Another lawsuit, this time instituted by a teacher named McLean against the Arkansas Board of Education, resulted in a State Supreme Court decision prohibiting the implementation of the act. Meanwhile, no less than eleven other states have debated, if not always introduced, legislation of a similar kind, and this has influenced the way in which the approach to evolution should be presented in schools.

The essential features of the anti-evolutionist doctrine for schools can best be summarised in the form of a series of statements:

(i) Creation of the Universe, matter, and energy occurred suddenly.
(ii) Creation of life occurred suddenly.
(iii) Creation of living kinds of plants and animals as they occur today occurred suddenly. Changes such as genetic variation only occur within the limits of the originally created forms.
(iv) The occurrence of mutation and natural selection is insufficient to have brought about the emergence of all living kinds from a single earlier organism.
(v) Man and apes have separate ancestries.
(vi) The Earth's geology is explicable in terms primarily of rapid catastrophic processes. (This was the view held by Cuvier in the early nineteenth century (see p. 21).)
(vii) Scientific dating processes indicate that the age of the Earth is much more recent than the several thousand million years postulated by evolutionists.

It is worth noting that the names of God and the Bible are not always included in writings on creation science but their influence is none the less implied.

## What is science?

In attempting to evaluate creation science and to determine whether its approach qualifies as truly scientific, we must first be clear as to exactly what we mean by 'science'. If we try to avoid philosophic side-issues the main features of science can be enumerated as follows:

(i) Science depends upon the existence of natural laws.
(ii) All scientific explanations are, ultimately, by reference to natural laws.
(iii) Scientific findings must be testable against empirical observation; that is, those that rely on experiment rather than theory.
(iv) As has been stressed by the philosopher Karl Popper, scientific theories cannot be proved for certain and must be susceptible of test. Their conclusions are essentially tentative and not final. Eventually, they may be replaced or modified as our knowledge advances.
(v) Experimental deductions can be proved to be false if and when a better interpretation is forthcoming.

## Is creation science scientific?

In order to decide whether creation science justifies the title of a 'science', we must attempt to contrast the characteristics of a truly scientific approach outlined above with those of a creationist one. The latter can be summarised as follows:

(i) Creationism depends upon a sudden creation from nothing, the agency concerned being supernatural intervention and not a natural law.
(ii) The explanation of the whole natural environment is by reference to divine creation and religious mysticism.
(iii) All deductions are, in reality, dogmatic assertions which cannot be tested or falsified.
(iv) Creationist conclusions, being dogmatic, leave no latitude for adjustment or alternative interpretation.

To make the comparison of science and creationism a little more realistic, it may be helpful to quote a few examples of typical statements made by creationists. These can be grouped under scientific and philosophical (religious) headings. Milne [7.6] has stated the scientific position clearly and quotes three typical creationist arguments:

(i) The low concentration of helium in the Earth's atmosphere is evidence of its relatively recent origin.
(ii) If estimates of the Earth's human population are projected backwards in time from the period AD 1650–1800, a date is reached (4320 BC) at which only two humans (presumably Adam and Eve) were present. This argument is similar to that of Archbishop Ussher in Britain (see p. 1).
(iii) The rate of oceanic sedimentation today could account for the entire deep-sea sediment layer in only 30 000 years.

Readers may care to judge for themselves the extent to which these claims are justified. If there are fallacies in the arguments, where do they lie?

On the philosophical and religious side, we are faced at the outset with a basic question namely, can God's will be used as part of a scientific explanation, as most creationists believe? Some of the attempts that have been made to equate biblical writings with accepted scientific facts have been ingenious, to say the least. As we saw earlier, a prime concern of creation science is to establish the literal truth of the origin of the Earth and life upon our planet, as recorded in the Book of Genesis. A major problem is how to relate the biblical duration of the Creation (6 days) with acceptable estimates of the extent of geological time. The so-called **gap theory** was devised some years ago and is periodically revived as one way of meeting the difficulty. This states that the original Earth described in Genesis 1:1 was populated by living things including 'pre-Adamic men', but was destroyed by a divine flood becoming 'without form and void' (8:2). There was then a protracted gap in time during which the fossils found in the Earth's crust today gradually accumulated. These represent the relics of the original perfect world destroyed before the 6 days of creation (or re-creation) recorded in Genesis 1:3–31, began. Perhaps it is only fair to add that many creationists reject the gap theory on the grounds that it provides a geological solution to the creation story, not a divine one. Space does not permit a more detailed consideration of similar ideas and the sort of arguments put forward to justify their acceptance. These are well documented and can be found in the appropriate literature [7.7].

It may be argued that speculative discussion of religious mysticism such as the gap theory have no place in a book purporting to be concerned with science. But it must be remembered that creation science in the United States has now accumulated an extensive literature, which has gained considerable publicity; moreover, there is an active 'research' organisation responsible for developing and promoting creationist doctrines both on the scientific and religious fronts. As we saw earlier, in some of the southern states the teaching of creation science in schools as a genuine alternative to traditional evolutionary theory is now required by law.

In attempting to evaluate creationist literature and utterances we are faced with procedures that deviate widely from those generally accepted as scientific. These include political manoeuvres, the misinterpretation, misunderstanding, or disregard of established scientific evidence, and a frequent recourse to religious philosophy and mysticism. However, as Milne [7.6] has admirably demonstrated, attempts at evaluation can be well worth while, particularly for students, as exercises in clear and critical scientific thinking.

# Darwinism and science

At the beginning of this chapter the point was made that Darwin's views on the origin of species were offensive to two sections of Victorian society – churchmen who held to a fundamentalist viewpoint and scientists who, if they accepted the idea of change, interpreted its occurrence in terms of the conventional thinking to which they adhered. For many objectors, the idea of evolution through natural selection was unacceptable on both grounds. Having considered some of the religious aspects, let us now turn to the scientific criticisms, which, like fundamentalism, still have a certain significance today.

### Buffon's explanation of change

Darwin's idea that plants and animals had changed throughout time was not new. The concept of change can be traced back more than 2000 years to the Greeks such as Aristotle and the Ionian philosophers of the sixth century BC. But by the eighteenth century AD, the popular view had become more static and the fixity of species was universally accepted. There were two main reasons for this outlook. The first, and by far the most important, was religious doctrine and a fundamentalist interpretation of the Bible. The second was an adherence to the idea of spontaneous generation – that living could arise from non-living matter. Thus, animals such as worms could be generated from the soil in which they lived, an assumption that was only finally disproved by the work of Louis Pasteur on bacteria in 1861.

The first signs of a change in outlook occurred among eighteenth-century French biologists. Georges Buffon is remembered largely for his monumental *Histoire Naturelle* in forty-four volumes, including a series of essays on a wide variety of topics. His outlook was typically that of a deist and he subscribed wholeheartedly to a fundamentalist viewpoint. However, he also accepted the idea of change, which he observed among the plants and animals that he studied and which he attributed to influences such as food and the physical environment. Thus, in discussing the origins of domesticated breeds of dogs he attributes their variety to the influence of the climate, the comforts of shelter, food, and careful training.

Buffon attempted a logical explanation of the origin of the Earth in terms of sedimentary rocks, pointing out that many species of animals had gone extinct before the arrival of man. In this he did for his distinguished successor Lamarck much what the geologist Lyell was to do for Darwin. But his contributions to evolutionary thinking were only fragmentary and no coherent theory emerged in his time.

## Lamarck's theory of evolution

Unlike Buffon, whose main preoccupation was with the facts of natural history as he observed them, his fellow countryman Lamarck took a more expansive view of the living universe. His ideas are to be found in the greatest of his works – *Philosophie Zoologique*, published in 1809. Central to his thinking was the idea of an evolutionary process in nature with a natural tendency towards increasing complexity. This had resulted in chains of life extending in a linear manner from the simple to the most complex organisms, plants and animals being separate from one another. His approach was essentially mechanistic; he regarded the distinction between the living and non-living as due merely to different levels of organisation.

Lamarck went further and propounded four laws governing the evolutionary sequence of living things. In summary, these are:

(i) Nature tends to increase the size of living individuals to a predetermined limit.
(ii) The production of a new organ results from a new need.
(iii) The development reached by organs is directly proportional to the extent to which they are used.
(iv) Everything acquired by the individual is transmitted to its offspring.

Laws (iii) and (iv) are of particular significance since they provide the basis for the differing versions of Lamarckism that survive today. The two commonest are the 'inheritance of the effects of use and disuse' and 'the inheritance of acquired characters (modifications)'. We must remember that Lamarck was unaware of the distinction between germ (reproductive) cells, which are transmitted from generation to generation, and somatic (body) cells, which are not. Their difference and its implications were finally established through the work of the distinguished German biologist Weismann at the time of Darwin about 50 years later. To Lamarck, the inheritance of modifications acquired during an organism's existence therefore presented no biological or philosophical difficulty. As to evolution, competition between organisms and their differential survival played no part in bringing it about, only their inherent capacity to become more complex and the effects of the environment in promoting inherited change. Although Lamarck was evidently a theist, adhering to the traditional view of a divine plan of life, he was, none the less, prepared to write that 'it has been thought that nature is God, indeed, it is the opinion of the majority . . .'.

That Lamarck's theory of evolution had a lasting impact in France there is no doubt. Thus, his extensive discourse on the origin of the long neck and legs of the giraffe acquired in response to the demands of browsing on trees was a subject of the cartoonist Caran d'Ache (Fig. 7.1) a hundred years later. In England, too, Lamarck had a considerable following and some of those who rejected *The Origin of Species* did so because the idea of natural selection conflicted with Lamarckian principles.

## Inheritance of acquired characters since Lamarck

Lamarck's views on inheritance were much the same as those of Erasmus Darwin, grandfather of Charles Darwin, put forward a few years earlier. Both believed that characteristics acquired during the lifetime of an organism could be transmitted to the next generation, but whereas Lamarck placed particular emphasis on the effects of use and disuse, Erasmus Darwin took a broader view, regarding the environment in general as being responsible for inherited change. As organisms became more complex, so they themselves influenced their environment which in turn influenced them further.

As we saw earlier (p. 142) one of the essential features of any scientific theory is that it must be capable of test and, if necessary, disproof. Lamarck's theory is, however, exceedingly difficult to test for several reasons:

(i) Supporters can always maintain that acquired modifications only take place gradually and to detect them requires a period of time far longer than any experiment would allow.

(ii) An experimental test would require the use of a group of organisms from which all genetic variation had been eliminated in order to ensure that any changes observed were of truly environmental origin. Such pure lines are difficult to produce and maintain.

(iii) Viewed in a modern context, a Lamarckian mechanism of change assumes the environmental control of the process of mutation. Now it is, of course, well known that mutation rate can be controlled by environmental influence such as ionising radiation. The point at issue is whether it can be directional in relation to a particular environmental requirement (i.e. adaptive).

Many attempts have been made in the past to establish the principles of Lamarckism. Notable among these were the claims by the Lysenko school of geneticists in Russia to have utilised a Lamarckian approach in the production of crops resistant to extreme environmental conditions. These assertions are difficult to evaluate in Western terms on account of their ideological overtones. However,

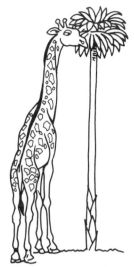

– Dis donc, papa, *pourquoi* que
les palmiers sont si grands?
– C'est pour que les girafes
puissent les manger, mon enfant,
car . . .

. . . si les palmiers étaient tout
petits, les girafes seraient très
embarrassées.

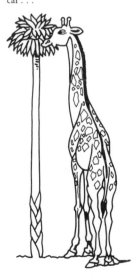

– Mais alors, papa, *pourquoi* que
les girafes ont le cou si long?
– Eh bien! c'est pour pouvoir
manger les palmiers, mon enfant,
car . . .

. . . si les girafes avaient le cou
court, elles seraient encore bien plus
embarrassées.

**Fig. 7.1** The problem of Lamarck's giraffe faced with the challenge of the environment (after the *Album de Caran d'Ache*, 'Les petits pourquoi de M. Toto')

two typical investigations have been well documented in Britain and these are useful to students in illustrating the kinds of problems and doubts that can arise in this problematical field of experimentation.

### Kammerer and the midwife toad

Paul Kammerer was a biologist working in Austria who committed suicide in 1926 after a protracted controversy over a series of experiments in which he claimed to have established the validity of the inheritance of acquired characters [7.8]. A full account has been provided by Koestler [7.9] and only a brief summary will be given here. The midwife toad, *Alytes obstetricans* (Fig. 7.2), unlike other toads, mates on land. The males lack the nuptial pads normally present on the forelimbs of other species to enable them to grip the slimy females in water. Kammerer forced *Alytes* to breed in water and claimed that after several generations, nuptial pads began to appear. From this he concluded that a modification derived from the environment was being inherited. During a visit to the laboratory the induced nuptial pads (whose usual colour is black) were examined and one was found to contain Indian ink. This led to his downfall. Were his other results genuine or were they, too, a forgery? We shall never know.

However, another of Kammerer's experiments (which he regarded as his most significant) involved amputation experiments on the sea-squirt, *Ciona intestinalis*, in order to induce regeneration and demonstrate acquired inheritance of increased siphon length. Unlike the investigations on *Alytes*, which have never been repeated, those on *Ciona* have been subjected to critical tests on several occasions, the latest being in 1975 as described by Ewer [7.10]. From these it seems extremely unlikely that the animals could ever have withstood the trauma of the kind of amputation described by Kammerer.

**Fig. 7.2** The midwife toad, *Alytes obstetricans*. The eggs are carried by the male until they hatch into tadpoles which then develop in water

The logical conclusion from Kammerer's attempts to demonstrate a Lamarckian effect in the midwife toad and other animals seems to be that many, if not all of his critical experiments, are not repeatable. That he knew what his results ought to have been there is little doubt; but to what extent they were fictional remains an open question. If, in fact, he did indulge in scientific deceit in order to establish preconceived ideas, his behaviour would not have been unique.

### Immunity and Lamarckism

The capacity for immune response is one of the most important endowments that animals possess, since it enables them to achieve protection from pathogenic organisms many of which would otherwise be fatal. The subject has received much attention in recent years and some of the systems conferring immunity have proved to be exceedingly complex. The following brief account of the work of Steele [7.11] and others is therefore treated only in the simplest terms. When the body is invaded by antigens, which are usually bacteria or viruses, antibodies are produced. Each antibody is specific to a particular antigen and is capable of neutralising and eventually eliminating it. One of the most remarkable things about the immune system is not only can it produce antibodies to a range of naturally occurring antigens, such as carbohydrates and proteins, but it also responds to substances synthesised by man, which neither the organism nor its ancestors can ever have experienced before. In other words, an animal can 'learn' to produce a specific antibody (protein) to almost anything! Since the code for protein production is located in DNA and RNA, it follows that antigen information must somehow be passed from the exterior environment to the hereditary system. The situation thus has a distinct Lamarckian ring about it. This is further supported by the now well-known phenomenon of **tolerance** in which individual organisms acquire the capacity for not regarding their own tissues as antigens and refraining from producing antibodies against them.

One of the problems in immunology is to account for the extraordinary diversity of immune responses. Some workers, including Steele [7.11], subscribe to the so-called **clonal selection theory**, which postulates that early in its development an animal produces many kinds of antigen-sensitive leucocytes each potentially capable of making a specific antibody (Fig. 7.3). It is argued that some antibody specificities that are species related, such as those for endemic disease antigens, are encoded in the DNA of the cell nucleus and inherited. The remainder arise by somatic mutation and are subject to subsequent selection during the life of the organism. A foreign antigen entering the body thus selects specific selector cells (leucocytes) more or less on the basis of the best fit. These then proceed to multiply and produce appropriate antibodies. The clonal selection theory appears to accommodate all the observed facts, including the elimination by selection of those leucocytes that would otherwise identify the cells of the organism as antigenic (tolerance).

The alternative **germline theory** holds that all antibody specificities are encoded in the genes of the animal and inherited *en bloc*. This implies that during the ancestry of the species a vast array of antigenic experiences must have occurred and the antibody potentialities accumulated through the process of natural selection. One of the main objections to this argument is that it does not explain the ability of the system to respond to antigens that it could never have experienced before.

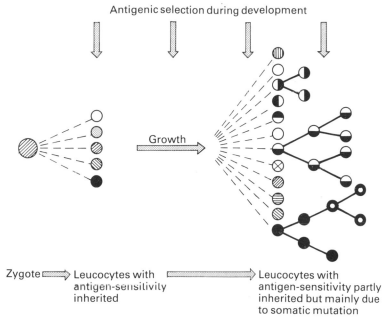

Antigenic selection during development

Growth

Zygote ⟹ Leucocytes with antigen-sensitivity inherited ⟹ Leucocytes with antigen-sensitivity partly inherited but mainly due to somatic mutation

**Fig. 7.3** Schematic representation of the clonal selection theory (after Steele 1979)

Gorczynski and Steele [7.12] carried out a series of experiments in which they injected foreign antigens in the form of a mixture of bone marrow and spleen cells at regular intervals into an inbred strain of mice. After acquiring tolerance, the male mice were then mated to normal inbred females of the same strain, the hypothesis being that immunity would be transmitted through the father. The breeding experiments appeared to support the claim for the independent inheritance of somatically acquired tolerance. In the first generation this reached the high level of 40–50 per cent, and in the second it was also well within the limits of statistical significance.

The procedures involved in this kind of experimentation are complex and not easy to replicate precisely. None the less, at least three attempts by workers experienced in the field of immunology have so far failed to reproduce the findings of Gorczynski and Steele. Their experiments were conducted under highly specific conditions involving a single kind of antigen and the paternal transmission of tolerance. If such acquired tolerance can, indeed, be transmitted in the circumstances of the experiment, are the conditions exceptional and inadequately understood? The fact remains that these experiments and others like them have not convincingly demonstrated the operation of Lamarckian effects. Thus, the verdict must be that the case is as yet unproven.

## Summary

1 Criticisms of Darwin and the views contained in *The Origin of Species* were on both religious and scientific grounds.
2 Some of the most vicious attacks were launched on behalf of the Church which was essentially fundamentalist in outlook, believing in the literal truth of the Bible.

3   The idea of the evolution of man from a primate stock proved particularly unacceptable to Victorians.

4   In his theory of evolution Darwin tended to overlook the origin of essentially human traits such as social values and the accumulation of knowledge. Development of these has been called psychosocial evolution. Whether this is a branch of biology seems uncertain.

5   Darwin's views on the origin of species were essentially materialistic. He saw no evidence of a divine plan for the development of the Earth involving a determinist approach.

6   The evolution of essentially human attributes such as self-awareness, ability to reason, and distinguishing good from evil is difficult to explain in terms of survival value alone. Does man have a 'soul' and, if so, should this be regarded as part of his biological makeup?

7   Altruism has played an important part in animal evolution, particularly in man. Some aspects of man's altruistic behaviour appear to lie outside the realm of natural selection.

8   In the United States, fundamentalism has been much more widespread than in Britain. Creation science is essentially dogmatic and cannot be evaluated through the usual processes of science. None the less, its influence in some quarters has been considerable.

9   Lamarck's theory of evolution represented the first attempt to put forward a coherent scientific explanation of change. It exerted a considerable influence against Darwinism. By its very nature, it is exceedingly difficult to disprove.

10  Kammerer's attempts to demonstrate Lamarckism using the midwife toad and sea-squirt failed because his experimental procedures and results proved to be suspect.

11  Attempts by Gorczynski and Steele to establish the operation of a Lamarckian mechanism in the acquisition and inheritance of immunity seemed promising. But so far their findings have not been confirmed.

## Topics for discussion

1   When, if ever, can we accept a theory as proved? What criteria can we use in deciding whether a hypothesis should be regarded as a theory?

2   It is claimed that life originated in its present form as a result of one or more acts of creation. What kinds of evidence would be necessary in order to support or refute this idea?

3   How would you rewrite the first chapter of the Book of Genesis so that it conformed to the neo-Darwinian theory of evolution?

4   It has been asserted that psychosocial evolution provides evidence for a Lamarckian explanation of change. To what extent can such a view be supported?

5   If Lamarck had been able to visit the Galápagos Islands, how might he have interpreted the adaptations among the tortoises and finches (see p. 41) that he would have found on the different islands?

6   Why is it likely that a widely held fundamentalist viewpoint, as in the southern United States, would eventually become linked to politics?

7   Construct an argument supporting the view that creationist science should be regarded as a true scientific discipline.

8 What evidence is there that man possesses some characteristics, such as certain kinds of altruism, which could not have evolved through natural selection?

9 Since Lamarck's explanation of change is impossible either to prove or to disprove, it should not be regarded as a genuine scientific theory. To what extent do you agree with this statement??

10 Does the statement that 'the meek . . . shall inherit the earth' (Matthew 5:5) conflict with modern ideas on natural selection?

# References

7.1 Bibby, C. (1972) *Scientist Extraordinary*. Pergamon

7.2 de Beer, Sir G. (1963) *Charles Darwin*. Nelson

7.3 Teilhard de Chardin, P. (1959) *The Phenomenon of Man*. Bernard Wall, London

7.4 Dawkins, R. (1976) *The Selfish Gene*. Oxford University Press

7.5 Mariner, J. L. (1977) 'The evolution–creation controversy in the United States'. *Journal of Biological Education* **11**, 6–11

7.6 Milne, D. H. (1981) 'How to debate with creationists – and "Win"'. *American Biology Teacher* **43**, 235–45

7.7 White, A. J. M. (1975) *What About Origins?* Dunestone Printers Ltd, Newton Abbot, Devon

7.8 Daws, J. A. (1977) 'Kammerer revisited'. *Journal of Biological Education* **11**, 21–6

7.9 Koestler, A. (1971) *The Case of the Midwife Toad*. Hutchinson

7.10 Ewer, D. W. (1977) 'Kammerer'. *Journal of Biological Education* **11**, 218

7.11 Steele, E. J. (1979) *Somatic Selection and Adaptive Evolution*. Croom Helm

7.12 Gorczynski, R. M. and Steele, E. J. (1981) 'Simultaneous yet independent inheritance of acquired tolerance to two distinct H–2 antigenic haplotype determinants in mice'. *Nature* **289**, 678–81

# 8

# Evolution: a modern synthesis

One of the merits of Darwin's theory as expounded in *The Origin of Species* was that, unlike any of its predecessors, it provided a workable explanation of evolutionary change that was susceptible of scientific test. Incidentally, it also provided a model of deductive scientific logic – an aspect seldom emphasised in textbooks of biology. But much has happened in the field of evolutionary biology since Darwin's time, and in this concluding chapter it is logical to ask where Darwinism stands today. Let us therefore examine the present state of thinking (sometimes referred to as **neo-Darwinism**) against the background of recent discoveries and the controversies they have provoked.

## The mode of inheritance

As we saw in Chapter 3, in attempting to formulate a coherent theory of the origin of species, Darwin was faced with problems that his critics were quick to exploit. Thus, on page 13 of the *Origin* he writes, 'The laws governing inheritance are for the most part unknown. No-one can say why the same peculiarity in different individuals of the same species, or in different species, is sometimes inherited and sometimes not so.' Against his better judgment, Darwin was obliged to accept the theory of his day. This held that inherited characteristics merged together at each generation thus reducing variation rather than promoting it. To offset the effect of blending and to provide the variance that the theory required, Darwin postulated a high rate of sporting (mutation), which conflicted with his observations in the field.

It is one of the ironies of biology that while *The Origin of Species* was being written, Mendel was working at Brno on his experiments with peas and laying the foundation of modern genetics based on the principles of the segregation and recombination of genes. In retrospect, it is indeed fortunate that Mendel happened to choose *Pisum sativum* as his experimental material and selected characters that were not subject to linkage and crossing-over; or to phenomena such as apomixis (setting of seed without fertilisation) that upset his later investigations of plants such as *Hieracium*. From an evolutionary standpoint, one of the most important realisations of recent years is that the beneficial effects of genes controlling such characters as reproductive capacity, hardiness, and protective coloration can come together (possibly by inversions) and can exist as closely linked groups (**supergenes**) that are virtually immune to crossing-over. As we now know, these provide the basis of many polymorphisms, such as those of the ABO blood group series which, by virtue of their selective advantages and disadvantages in different circumstances, have played an important part in evolution (p. 116).

## The nature of variation

Spectacular advances in our knowledge of the biochemistry of the cell, in particular of the nature of DNA, RNA, and the triplet genetic code, have thrown much new light on the origin of variation. Aided by techniques such as electrophoresis, it has been shown that mutation at the biochemical level may be more diverse and can occur at a much higher rate than had previously been envisaged. The selective advantage and disadvantage (if any) of particular biochemical mutants are still far from clear; nor is it certain that those mutants are subject to natural selection in the usual Darwinian sense. If, indeed, a proportion of them are of neutral survival value, it could be argued that random gene survival (**genetic drift**, p. 104) is a more significant factor in promoting change at the molecular level than it is in populations of whole organisms.

Modern discoveries have also extended to the nature of the gene itself. Although it is generally accepted that each gene is responsible for a particular protein (or part of one), it is now known that at least in higher organisms, sequences of nucleotides coding for particular proteins (**exons**) alternate with inert sections (**introns**), which have no effect and are lost at the formation of messenger RNA. What these introns represent in evolutionary terms is uncertain; they could be portions of supergenes that for unknown reasons have lost their function. Another aspect of supergenes relates to the finding that genes may duplicate one another, as has happened in the gene controlling the $\gamma$ chain of the haemoglobin molecule, which has evidently arisen by duplication of the gene for $\beta$ haemoglobin. This process of gene multiplication contrasts with that of chromosome fragmentation (p. 63) and provides a further source of new variation.

But in spite of modern advances in our knowledge of gene chemistry, the fundamental question remains – is the total fund of variance among living organisms sufficient to account for the evolution of life by a neo-Darwinian proteins? Hoyle [8.1] has stressed the extreme improbability that the 200 000 proteins on which we depend could have arisen at random. He likens the probability of such an occurrence to the throwing of dice and the appearance of five million sixes in succession. But this is to disregard an important part of the selectionist argument, which is that the evolution of biochemical (and other) systems over time has *not* been a random affair. Any mutation with even a slightly beneficial effect will tend to be selected positively and eventually to become established in the constitution of the cell. Should a similar variation arise later, it will have the effect of reinforcing a change already begun, the result being cumulative. Such a mechanism could have the effect of greatly speeding up the development of more complex chemical substances. Moreover, the process is likely to be enhanced by the situation mentioned earlier, that mutation at the biochemical level may be at a considerably higher level than elsewhere.

In attempting to explain the origin of the existing range of variation, Hoyle has postulated the need for an external source of genetic material. He believes that this material showers down on the Earth's surface in the form of cosmic particles. The DNA these particles contain has then become incorporated with that of microorganisms and other existing life-forms. One difficulty with this theory is that evidence of such genetic fusion is lacking, nor is any mechanism known whereby extraneous matter of environmental origin could be incorporated in the genotype of an existing organism.

There is, however, another way in which cosmic influences could have exerted significant effects on variation. As we saw earlier (p. 65), the rate of mutation can be greatly accelerated by mutagenic agents, some of the most powerful being ionising radiations such as occur in the atmosphere. Today, the intensity of cosmic radiation striking the Earth's surface is comparatively low but, in the past, this could well have fluctuated in magnitude with corresponding effects on the fund of inherited variation. Unfortunately, as in so many other aspects of evolution, precise evidence of conditions in the past is still lacking.

## Natural selection

Darwin never studied natural selection in action; he only deduced that it must take place. This theoretical conclusion was reinforced by his observation of adaptations among plants and animals in nature; in particular to the peculiar ecological conditions in the Galápagos Islands (p. 41). Presumably these adaptations were the outcome of a selective process working on variable species. The nearest he ever came to seeing the operation of this process was in the field of domesticated species such as pigeons, with which he was well acquainted and to which he devoted the first chapter of *The Origin of Species*.

An outstanding feature of the post-Darwinian era has been the development of a branch of study known as **ecological genetics**. This covers all aspects of the adaptations of wild populations to their environment and employs appropriate mathematical methods of sampling and analysis. One of the problems of studying evolution experimentally is that it is essentially a slow process. To achieve worthwhile results, it is therefore necessary to find situations where rapid changes are likely to occur such as:

(i) Populations that fluctuate a great deal so promoting outbursts of variation available for selection.

(ii) Characteristics that are under polygenic control and show continuous variation in breeding populations (**demes**) either isolated by physical barriers (**allopatry**) or subject to powerful selection pressures that serve as an isolating mechanism (**sympatry**).

(iii) Species invading a new habitat such as an island.

(iv) Situations involving genetic polymorphism (p. 83).

One of the striking outcomes of such studies has been the realisation that natural selection is a far more powerful process than had previously been supposed. Selective advantages of one variant over another of the order of 30 per cent and more are quite common, and in some species can be regarded as the norm. This means that appreciable changes in the gene frequency of populations (i.e. microevolution) in response to environmental fluctuations can be detected and assessed in a matter of only a few generations. Such findings have also been useful in enabling us to assess more precisely the likely contribution of random (non-selective) processes such as genetic drift to evolutionary change. That this occurs in all populations there is no doubt. But bearing in mind the magnitude of natural selection, the chances that drift will exert an appreciable effect on gene frequency are remote, even in the smallest viable populations.

## Is evolution directional?

One of the contemporary objections levelled at Darwin's theory was that a concept of change based upon variation and natural selection failed to explain the apparent tendency among some plant and animal groups to evolve in particular directions, suggesting that their evolution was somehow directed towards a particular goal. The idea of a directional urge in the development of life was not new. Part of Lamarck's theory of evolution (p. 144) had been that all organisms possess an innate urge to increase in complexity, an idea which Darwin rightly rejected. On a broader plane was the supposition that the evolution of certain groups such as the horse towards increasing size and a gradual reduction in the number of toes was somehow predetermined. The tendency for related groups of organisms to evolve in the same direction was known as **orthogenesis**.

The idea of orthogenesis has proved particularly attractive to fundamentalists and creationists, who have seen in it the hand of the Creator and the un-folding of a divine plan. One of the reasons why belief in orthogenetic patterns of change was so prevalent in Victorian times was the inadequacy of the fossil record. We know now that nearly all instances of apparently linear evolution, including that of the horse, have been associated with much lateral branching and adaptive radiation into a variety of ecological situations. Many of these side branches soon became extinct but some survived for considerable periods in various parts of the world.

It could be argued that directional evolution is precisely what we should expect as an outcome of a neo-Darwinian mechanism. Once a particular variation has proved beneficial in a given set of conditions and become established, the chances are greatly increased that further development will take place in the same direction. This is not, of course, because mutation is directional but because a similar mutation or combination of genes, when it occurs, will have the effect of enhancing a process already begun. Similarly, variations in different directions will have a correspondingly reduced chance of survival. The process of progress-ive selection will be further accentuated by the nature of environmental changes, which, in the absence of upsets by human interference, tend to move from one condition to another in an orderly fashion rather than violently fluctuating.

Another aspect of directional evolution arises from situations in which organisms that are only distantly related are faced with the same environmental problems, such as the need to respond to the stimulus of light. Similar visual mechanisms can then be evolved but in quite different ways. Such struc-tures therefore exhibit **analogy** rather than homology and are examples of **convergence** in evolution. As we have seen, distinguishing between true homologous characteristics and analogous resemblances resulting from conver-gent evolution is not always easy, and can lead to considerable difficulties in establishing relationships at the level of the higher taxonomic groups (orders and above). A typical example is provided by the evolution of a type of eye such as our own, consisting of a lens that focuses incoming light onto a sensitive surface (retina) connected to an elaborate nerve supply. Such an arrangement has been evolved many times over and reached varying degrees of elaboration among different groups of invertebrates. The climax is reached among cephalopod molluscs (squids and cuttlefish), where the structure of the eye bears an extraordinary resemblance to that of mammals (Fig. 8.1). However, closer examination of detail, for instance the distribution of the cells of the retina, shows that the two eyes are formed in quite different ways and that their similarities are due to convergent evolution and do not indicate close lineage. A comparable example concerns the evolution of wings. Although all kinds have a common function, those of vertebrates such as birds bear no resemblance to the comparable structures occurring in insects.

(a)

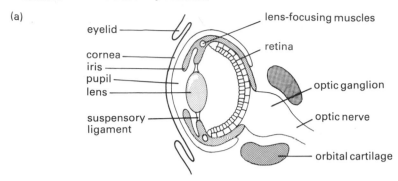

(b)

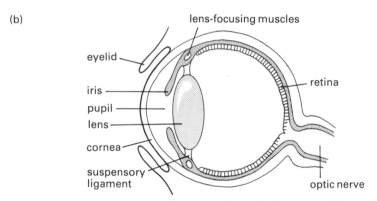

**Fig. 8.1** Convergent evolution: (a) diagrammatic vertical section of the eye of a cephalopod mollusc (squid), (b) a similar view of the eye of a mammal

## Evolution: continuous or discontinuous?

One of the greatest advances in evolutionary biology since Darwin's time has been an increase in our knowledge worldwide of the fossil record. This has served to substantiate Darwin's conclusion that the rates of change among living things have varied greatly in different places and at different times. It has also highlighted the discoveries of Cuvier, made among the fossil beds near Paris (p. 21), that discontinuities can occur because species found in one stratum can be absent from another nearby. Many such instances are now known, notable among them being the gastropod and bivalve molluscs found in the Pliocene and Pleistocene deposits of the Turkana Basin in northern Kenya (p. 37). This has led to the idea that evolution has not been a gradual process but has consisted of periods of comparative stability alternating with outbursts of variation and the formation of new species. This is now known as **punctuated** (or **punctuational**) evolution. As might be expected, the duration of periods of stasis and outbursts of evolutionary activity vary greatly in different circumstances.

But not all evolutionary change is of the discontinuous kind. Lande [8.2] has argued that evidence from a range of living genera of lizards compared with that from fossil sequences suggests that a transition from the typically lizard-like form with complete limbs to a limbless, snake-like body must have been gradual and covered many millions of years. Thus the fundamental difference between continuity and discontinuity is to some extent a matter of degree.

The important question, however, is not only what variations occur in the geological time-scale of species but whether a single mechanism (neo-Darwinism) can be held responsible for all the patterns of evolution that they reveal. Moreover, can a process regarded as effective for individuals and species also be responsible for the degrees of divergence that characterise the higher taxonomic levels?

## Neo-Darwinism in context

The essence of the neo-Darwinian argument is that the adaptation of organisms to their environment is brought about by natural selection acting upon small inherited variations, most of which are initially non-adaptive. Mutation is the ultimate source of new variation and it is preserved and transmitted by the mechanism of particulate inheritance. Changes in the gene frequencies of different populations (in microevolution) leading to the formation of races and, eventually, new species depend upon various forms of reproductive isolation and the restriction of gene flow. Although selection pressures are often severe, the formation of new species is essentially a slow process.

An objection to neo-Darwinism by the punctuationists is that it fails to account for discontinuity and the outbursts of rapid evolutionary activity that have alternated with those of inactivity. The latter could, of course, be attributed to periods of ecological stability alternating with those of rapid environmental change. But the inherent slowness of a neo-Darwinian mechanism has prompted the idea that macromutations involving much larger phenotypic changes may also have played a part. On theoretical grounds it could be argued that major changes either at the sites of genes (**point mutations**) or in the structure of chromosomes would be likely to cause gene incompatibility and hence to upset the delicately balanced gene complex. Certainly, few such changes have been observed; and we have the example of polyploidy to illustrate how spontaneous alterations in chromosome numbers usually result in sterility.

Allied to the idea of sudden major changes in the evolutionary pattern is the problem posed by intermediate stages. Such stages must have preceded many adaptations that eventually proved successful. For instance, the lens-type eye quoted earlier (p. 155) was presumably preceded by a range of devices that concentrated light in varying degree but did not form an image. Could their possessors have been at a selective disadvantage on account of such semi-adapted structures? If so, how did the lens–retina mechanism manage to survive? We can only guess at possible answers. However, any structure that increased the capacity of an animal to perceive light of varying intensity could have been an advantage in the detection of shadow and movement, and hence in aiding survival from predators and the identification of an optimum habitat. We need go no further than the phylum Mollusca, to find just such a sequence of evolutionary stages still existing today (Fig. 8.2).

The fact that a structure performs a particular function at one stage of its evolutionary history does not necessarily imply that its role has always remained the same. Thus, the swimbladder of fishes is a hydrostatic organ that enables the animals to cruise and feed at a particular depth. But in the lungfishes (order Dipnoi), it has acquired the respiratory function of a lung in response to the need to survive periods of desiccation and oxygen shortage in the dried-up mud of river beds. The existence of a swimbladder in the ancestors of the lung fishes, no matter

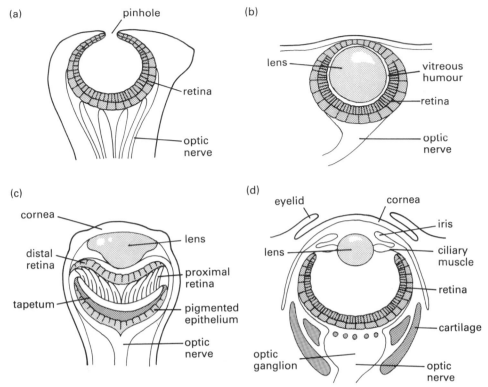

**Fig. 8.2** Eyes of molluscs: (a) *Nautilus* (cephalopod), pin-hole eye with retina but no lens; (b) snail, *Helix* (gastropod), fixed lens and simple retina; (c) scallop, *Pecten* (bivalve), fixed lens and complex retina; (d) cuttlefish, *Sepia* (cephalopod), movable lens and complex retina (from Kerkut 1961)

how rudimentary, was therefore a **preadaptation** for other ecological requirements.

Similarly, the feathers of the ancestral birds before *Archaeopteryx* (p. 30) may well have evolved as a means of retaining heat and stabilising the body temperature to a condition of homoiothermy. Only later did they acquire the additional function of aiding flight for which they were already preadapted.

It seems likely, therefore, that preadaptation may have played an important part in the evolution of many organisms – plants and animals – helping to determine the extent of their eventual divergence into higher taxonomic groupings such as families, orders, and classes. If this is so, it detracts from the need to postulate a role for macromutations.

## Species selection

Contrary to the arguments outlined above, some supporters of a punctuational approach to evolution (e.g. Stanley [8.3]) have distinguished between selection and change occurring among **individuals** (microevolution) and that taking place among species and above (macroevolution). They have pointed out that natural selection is a process whereby those individuals survive that are able to reproduce more successfully than others. It can therefore only exert its effects in living populations. But the history of evolution as recorded in the rocks abounds with instances where whole species have gone extinct; indeed, over time, far more have failed than survive today. Since extinction involves eventual failure, it follows that no individuals survive to reproduce. Hence, the species rather than the individual is the unit of selection, leading to macroevolution rather than microevolution. In general, it is believed that some species, subject to a degree of isolation, tend to survive at the expense of others because they are better adjusted as groups of populations to the prevailing physical and biotic conditions and, as a result, they are able to diversify more readily into new species. But the kinds of factors promoting such changes are the same as those invoked by the neo-Darwinists – physical conditions such as climatic variation and biotic factors including predation and interspecific competition.

## Gradualist or punctuationist?

The essential difference between the gradualist and punctuational approaches is that the thinking of the first is vertical whereas that of the second is horizontal. Thus, on a gradualistic model, the rate of formation of higher taxa such as families and orders will be proportional to the time (as measured by the number of generations) over which divergence has occurred. On a punctuational model, the range of diversification will be proportional to the amount of splitting. Calow [8.4] has quoted numerous examples illustrating these principles and the relationship between time and rate of divergence. Thus, speciation in the evolution of mammals was rapid, with radiation extending to around 100 families in 30 million years. By contrast, the bivalve molluscs have evolved much more slowly; it required about 500 million years for them to attain roughly the same level of diversity as the mammals.

It follows from this argument that in groups of organisms where evolutionary changes have been slow there should be little speciation. This appears to be true for several groups of animals such as the lungfishes. If we assume that the rate of appearance of new species is proportional to the rate of morphological change, then the work of T. S. Westoll provides us with useful information on lungfish divergence. He measured twenty-six different characters such as the shape of the skull and nature of dentition, all of which were assigned numerical grades. Each fossil genus was then scored according to the degree of difference from a supposed ancestor, which was given a score of zero. Total scores ranged from 4 in the earliest genus to 100 for the two genera alive today. Plotting these scores against time gives the graph in Fig. 8.3. It will be seen that from the mid-Palaeozoic, morphological change was rapid for about the first 100 million years. Thereafter, it slowed down abruptly and for the last 150 million years hardly any changes took place.

On the face of it, there appears to be considerable evidence in support of a punctuationist interpretation of speciation. However, we must be cautious in assuming its universal operation from the data provided by a limited number of examples. A major problem in comparing the two viewpoints is to ensure that the criteria used in determining the higher taxa are reasonably comparable. As we have seen (p. 38), the approach to taxonomy of geologists and biologists is bound to be different. The only evidence provided by fossils relates to the preserved hard structures. Little or nothing is known about soft parts, modes of behaviour, or of the ecological conditions in which extinct organisms lived. The meaning of

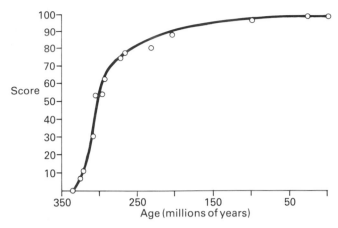

**Fig. 8.3**  Rate of evolution in lungfishes. Each point represents the score for a single genus. The lowest value is the most primitive condition, the highest the most modern one (after Westoll and Simpson 1949)

'speciation' to a geologist and a biologist may therefore be quite different. A criticism sometimes levelled at neo-Darwinism is that it is unable to predict a discontinuous pattern of change. But this is to misunderstand the mode of action of natural selection; for there is no reason why selection pressures should not vary, being intense during periods of environmental fluctuation but less so in conditions of environmental stability. The evolutionary history of the lungfishes (Fig. 8.3) also illustrates an important aspect that is often overlooked by enthusiasts for punctuationism when referring to 'outbursts' of speciation. It will be seen that the maximum radiation took place during the early period of evolution but, even so, this covered an immense time span of around 25 million years. Now that it is realised that selection pressures in nature frequently attain a level of 30 per cent or more, it could reasonably be argued that the length of time was ample for the formation of the observed number of species by a neo-Darwinian process.

The verdict in favour of one mechanism or the other must therefore remain uncertain. Such has been the heat of the debate that the two explanations have tended to be seen as mutually exclusive alternatives. In fact, there would seem to be no good reason why both should not have contributed in varying degree to the taxonomic diversity of plants and animals that we find today. Certainly, the discontinuity of some fossil series and the continuity of others suggest that this could be so. Only time and further research will tell.

## Taxonomic systems

The controversy raised by cladism (p. 121) and its merits relative to other taxonomic systems have served to highlight certain features of taxonomy in general. First, it is a system invented by man for bringing order out of chaos among living and extinct organisms. There is nothing new in this idea; it was fully appreciated by Aristotle in the fourth century BC. What the early Greeks failed to realise and only became evident in the eighteenth century, was the existence of homology – that many structures such as five-fingered limbs have a fundamental organisation in common, suggesting evolutionary affinity.

The essential feature of any natural classificatory system, be it cladistics, phenetics, or phylogenetic trees, is that all are founded on homologous resemblances. Hence, in their different ways, they reflect patterns of evolution. For any group of organisms there can be only one true phylogeny, but the same cannot be said for the corresponding taxonomy, because this is man-made. A valid criterion for deciding which taxonomic system to adopt must therefore be the extent to which it meets our particular needs. In the previous pages all three systems have been used on occasions to fulfil specific and differing requirements.

## What is fitness?

One of Darwin's greatest contributions was to point out that individuals and species survive and change as a result of the action of natural selection. This theme has permeated this book and it is therefore an appropriate one on which to end. Unfortunately the misplaced enthusiasm of one of Darwin's friends, Herbert Spencer, resulted in the coining of the phrase 'survival of the fittest'. This was judged to be more explicit than 'natural selection'. Although it was never used by Darwin himself, it managed to find its way into the later editions of *The Origin of Species* and it did much to confuse subsequent thinking. In the nineteenth century, the idea of a 'struggle for existence' was not new; many authors before Darwin had discussed it, including the Roman poet Lucretius in the first century BC. If we merely equate fittest with healthiest, it would be quite possible for the 'fittest' to survive without any evolution taking place. One of the shortcomings of early Darwinism was that it tended to emphasise survival whereas the role of natural selection, in reality, is to favour those parents capable of producing the most offspring.

The neo-Darwinian approach to natural selection lays far more emphasis on factors that influence differential reproduction within populations of species. Ultimately, population density depends upon a balance of births, deaths, immigration, and emigration. Each of these is, in turn, governed by the level of adaptation achieved by plants and animals to their physical and biotic environments. That this is a prime concern both of the ecologist and of the evolutionist only serves to underline the point made in the preface that a study of ecology can provide the most realistic point of entry into the study of evolution.

## Summary

1  One of the merits of Darwin's theory of the origin of species was that it proposed a mechanism of evolution that was susceptible of scientific test.

2  Mendel's discovery of particulate inheritance and its subsequent elaboration have provided an explanation of the transmission of variation – a problem that Darwin never solved.

3  Discoveries in cell biochemistry, particularly of DNA and RNA, have done much to explain the nature of the gene, also of gene mutations as the ultimate source of new inherited variation.

4  The field of ecological genetics has thrown much new light on the nature of natural selection and the magnitude of selective advantages and disadvantages.

5  There is no evidence to support the idea of orthogenesis (evolution in a predetermined direction). However, the neo-Darwinian model predicts that

once evolution has set off in a particular direction it is likely to continue along similar lines.

6 Convergent evolution between two distantly related groups of organisms can lead to the development of apparently similar structures. However, more detailed study shows them to be only analogous.

7 The fossil record shows that in a number of groups of plants and animals, evolution has been discontinuous (punctuated), periods of evolutionary activity alternating with periods of comparative stasis.

8 Macromutations may have been invoked to explain major evolutionary changes but there is little evidence to support the idea. The preadaptation of structures for purposes other than their immediate use has probably played an important part in evolution.

9 It has been suggested that at the level of higher taxa (families and above) the species rather than the individual is the unit of selection. This is a central assumption of the punctuational approach to evolution.

10 Although there is much fossil evidence to support a punctuational viewpoint, the data could also be explained satisfactorily along neo-Darwinian lines. It could be that both systems have operated in varying degrees.

11 There can be only one true phylogeny for any group of organisms but several taxonomic systems, depending on human requirements. Cladistics, phenetics, and phylogenetic trees all have their advantages and limitations.

12 Darwin's view of fitness was orientated towards success in the struggle for survival. The neo-Darwinian approach emphasises adaptation leading to increased reproductive success.

## Topics for discussion

1 The theory of evolution has so changed since Darwin's time that it should now be referred to not as neo-Darwinism but post-Darwinism. To what extent is there justification for this view?

2 What is meant by 'fitness' in the context of natural selection and how have views changed since Darwin?

3 How, if at all, does natural selection affect human populations today?

4 The phenomenon of preadaptation in organisms has been used as an argument for directional evolution (orthogenesis). What justification is there for this claim?

5 How do Darwinism and neo-Darwinism differ?

6 How does species selection (proposed by supporters of a punctuational approach to evolution) differ from individual selection (advocated by neo-Darwinists)?

7 What are the relative advantages and disadvantages of cladistics, phenetics, and phylogenetic trees as means of showing taxonomic relationships?

## References

8.1 Hoyle, Sir F. (1983) *The Controversy of Creationism*. British Broadcasting Corporation

8.2 Lande, R. (1982) 'Microevolution in relation to macroevolution'. In *Evolution Now: a Century after Darwin* (ed. J. Maynard Smith). Nature/Macmillan

8.3 Stanley, S. M. (1981) *The New Evolutionary Timetable*. Basic Books, New York

8.4 Calow, P. (1983) *Evolutionary Principles*. Blackie

# Bibliography

The following list is only a small selection of the many books written recently and in the past on different aspects of evolution. Most of the works included here are at a level approximately the same as this book. More advanced accounts, which may be useful for reference, are indicated by an asterisk.

## General evolution

Berry, R. J. (1982) *Neo-Darwinism* (Studies in Biology No. 144). Edward Arnold. (A brief account of the post-Darwinian theory of evolution including a chapter on other theories)

British Museum (Natural History) (1980) *The Origin of Species*. Cambridge University Press. (Well-illustrated pictorial guide to the British Museum's exhibit on evolution)

Calow, P. (1983) *Evolutionary Principles*. Blackie. (Concise and well-written account of the modern theory of evolution)

Darwin, C. (1859) *The Origin of Species*. Numerous reprints available, e.g. 6th edn. Penguin

*Dobzhansky, T., Ayala, F. J., Stebbins, G. L. and Valentine, J. W. (1977) *Evolution*. W. H. Freeman, San Francisco. (An advanced and comprehensive account of modern ideas)

Dowdeswell, W. H. (1975) *The Mechanism of Evolution*, 4th edn. Heinemann. (Emphasis on microevolution and the action of natural selection)

Gould, S. J. (1978) *Ever Since Darwin*. Burnett Books. (A collection of stimulating essays on evolution)

Hutchinson, P. (1974) *Evolution Explained*. David and Charles. (Good elementary introduction)

Maynard Smith, J. ed. (1982) *Evolution Now: a Century after Darwin*. Nature/Macmillan. (A provocative collection of essays covering the most important present-day issues in evolutionary biology)

Maynard Smith, J. (1975) *The Theory of Evolution*, 3rd edn. Penguin. (Evolutionary theory clearly explained)

Stanley, S. M. (1981) *The New Evolutionary Timetable: Fossils, Genes, and the Origin of Species*. Basic Books, New York. (The argument for a punctuational view of evolution)

Thornton, I. (1971) *Darwin's Islands: A Natural History of the Galápagos*. Natural History Press, New York. (The Galápagos Islands set in the context of Darwin's visit)

## Genetics and evolution

Berry, R. J. (1977) *Inheritance and Natural History*. Collins. (A 'New Naturalist' volume relating genetics to the countryside)

British Museum (Natural History) (1980) *Man's Place in Evolution*. Cambridge University Press. (An excellent short introduction, beautifully illustrated)

Dawkins, R. (1976) *The Selfish Gene*. Oxford University Press. (Evolution and genetics seen from the aspect of the gene)

*Ford, E. B. (1975) *Ecological Genetics*, 4th edn. Chapman and Hall. (Advanced text of ecological genetics)

Ford, E. B. (1981) *Taking Genetics into the Countryside*. Weidenfeld and Nicolson. (A stimulating account of the relationship between genetics and ecology)

Ford, E. B. (1979) *Understanding Genetics*. Faber. (An elementary introduction to the subject)

*Lewontin, R. C. (1974) *The Genetic Basis of Evolutionary Change*. Columbia University Press, New York. (Advanced account with an emphasis on population genetics)

*Murray, J. (1972) *Genetic Diversity in Natural Selection*. John Murray. (Variation in organisms and the action of natural selection)

## Human evolution

Aiello, L. (1982) *Discovering the Origins of Mankind*. Longmans

*Campbell, B. G. (1974) *Human Evolution*, 2nd edn. Heinemann. (A well-known work of reference)

Gribbin, J. and Cherfas, J. (1982) *The Monkey Puzzle*. Bodley Head. (Lively account of modern views on human evolution, raising a number of controversial issues)

Leakey, R. (1981) *The Making of Mankind*. Michael Joseph

Mather, K. (1966) *Human Diversity*. Oliver and Boyd. (A brief study of human variation and the action of natural selection upon it)

Pfeiffer, J. (1978) *The Emergence of Man*, 3rd edn. Harper and Row

Reader, J. (1981) *Missing Links*. Collins

## Darwin

de Beer, Sir G. (1963) *Charles Darwin: Evolution by Natural Selection*. Thomas Nelson (reprinted by Greenwood Press, 1976). (Excellent scientific biography)

George, W. (1982) *Darwin*. Fontana

Howard, J. (1982) *Darwin*. Oxford University Press. (Well-written short biography)

Vorzimmer, P. J. (1972) *Charles Darwin: the Years of Controversy*. University of London Press. (A detailed review of *The Origin of Species* and its critics, 1859–82)

## Origin of life

Folsome, C. E. (1979) *The Origin of Life*. W. H. Freeman, San Francisco. (Clear and concise account of modern views)

*Miller, S. L. and Orgel, L. E. (1974) *The Origin of Life on Earth*. Prentice-Hall, Englewood Cliffs, N.J. (Good synthesis of modern ideas)

Watson, J. D. (1980) *The Double Helix*, 2nd edn. Penguin. (The story of the discovery of DNA)

# Taxonomy

Crowson, K. A. (1970) *Classification and Biology*. Heinemann. (A good general account of taxonomic principles)

*Mayr, E. (1942) *Systematics and the Origin of Species*. Dover Publications, New York. (A classic study of systematics as seen by a zoologist)

*Mayr, E. (1970) *Populations, Species and Evolution*. Harvard University Press, Cambridge, Mass. (A general view of systematics in relation to evolutionary change)

*Sneath, P. H. A. and Sokal, R. R. (1973) *Numerical Taxonomy*. W. H. Freeman, San Francisco. (An authoritative account by two of the main advocates of numerical taxonomy)

# Creationism

Boardman, W. W. (1971) *Worlds Without End*. Creation-Science Research Center, San Diego, California. (Publication by the central body in the USA for promulgating creationist ideas)

Gillespie, C. (1978) *Charles Darwin and the Problem of Creation*. University of Chicago Press, Chicago. (Wider context of nineteenth-century Darwinism)

White, A. J. M. (1975) *What About Origins?* Dunestone Printers Ltd, Newton Abbot, Devon. (A typical example of the creationist approach to evolution)

# Glossary

**Abiotic factors**  Ecological factors of a non-living (i.e. physical) kind

**Adaptation**  Adjustment to a particular set of ecological conditions

**Adaptive radiation**  The evolution of a related group of organisms along a number of different lines involving adaptation to a variety of ecological conditions

**Alleles**  Alternative forms of a gene occupying identical positions on homologous chromosomes

**Allopatric evolution**  Evolution of a group of organisms following their isolation from the parent stock by a physical barrier (see **sympatric evolution**)

**Allopolyploid**  A polyploid organism resulting from chromosome multiplication in a hybrid

**Altruism**  Unselfish behaviour for the benefit of others of the same species; in an extreme form it may include self-destruction

**Analogy**  Resemblance of organisms to one another relating to the functions of their organs but not their structure

**Aneuploids**  Chromosome numbers in a species resulting from polyploidy or polysomy which are not simple multiples of one another

**Apomixis**  The formation of seeds in flowering plants without the fusion of gametes

**Aposematic colours**  Those that advertise the presence of an organism to potential predators to denote the inadvisability of attack

**Area effects**  Patterns of variation among populations of a species that are unrelated to any observable selective agents

**Artificial classification**  A taxonomic system based on analogy only

**Autopolyploid**  A polyploid organism resulting from the failure of the chromosomes to separate at mitosis of the zygote

**Balanced polymorphism**  A situation in which two or more forms of a polymorphic species achieve stability of numbers (see **transient polymorphism**)

**Binomial nomenclature**  The system by which every organism is given two names, indicating the genus and species to which it belongs; the genus is always prefixed with a capital and the species with a small letter

**Biometry**  Genetic studies involving the characteristics of families or organisms, and the correlation of different sets of variables

**Biotic factors**  Biological factors resulting from the interaction of living organisms with one another, such as food and population density

**Chloroplasts**  The structures in the cytoplasm in which chlorophyll is contained in green plants; their general composition resembles that of mitochondria

**Chloroplast inheritance**  A kind of inheritance among plants in which the egg only is responsible for transmitting degrees of greenness; the pollen makes no contribution

**Chromosome** A thread-like nuclear body of varying size and shape which carries the genes. The number of chromosomes per nucleus is characteristic for each species

**Chronospecies** A concept of species used by geologists involving statistical parameters only, such as shape and size

**Clade** The basic taxon of cladistics; organisms grouped within it are assumed to be derived from the same ancestor

**Cladistics** A school of taxonomy first formulated by the German biologist W. Hennig and based on the principle of genealogy

**Cladogram** A cladistic diagram showing the supposed relationship between different clades and based on shared characteristics that are considered homologous

**Class** A group of similar orders

**Cline** Graded variation within a species, either structural or physiological, taking place within a particular zone

**Clonal selection** A theory of immune response based on the idea that, early in its development, an animal produces many kinds of antigen-sensitive leucocytes each potentially capable of making a specific antibody

**Clone** A group of organisms or cells derived from a single parent and therefore genetically identical

**Coacervates** Cell-size accumulations of colloidal particles such as proteins produced artificially by adjusting the conditions of the environment such as the pH

**Coevolution** A situation in which plants and animals have evolved together in close interrelationship with one another

**Colchicine** A substance extracted from the meadow saffron, *Colchicum autumnale*, which acts as a mutagen by inhibiting the formation of the spindle at mitosis causing a doubling of the chromosome number

**Continuous variation** Variation occurring as a graded series, such as human hair colour

**Convergent evolution** A situation in which two or more different groups of organisms have come to resemble one another in evolution, but where there is no evidence of a common ancestry; resemblances are therefore analogous rather than homologous

**Creationist science** Fundamentalist beliefs based on the idea of creation and a rejection of organic evolution; prevalent in southern United States

**Crossing-over** An interchange of groups of genes between the members of a homologous pair of chromosomes

**Cultivar** An agricultural or horticultural variety of a plant species

**Cyanogenesis** The production by a plant of the chemical hydrogen cyanide

**Cytoplasmic inheritance** The inheritance of characters transmitted by organelles other than the cell nucleus

**Deletion** A kind of chromosome fragmentation (mutation) involving the removal of a block of genes

**Deme** Part of a population representing a distinct breeding group

**Density-dependent factors** Ecological factors such as food whose influence is governed by the density of the population on which they act

**Derived characters** A term used in cladistics to define characters that are thought to be genuinely homologous in different species

**Diploid**   The condition in a cell or organism when both representatives of each homologous pair of chromosomes are present

**Directional selection**   Selection of a variant within a species giving a skew distribution in favour of the variant

**Discontinuity**   A situation in which a particular group of fossils occurs in a geological stratum but not in an adjacent one

**Discontinuous variation**   Variation in a character which assumes several distinct forms that do not form a graded series

**Disruptive selection**   A situation in which a quantitative variant is at a selective advantage at its two extremes but at a relative disadvantage in the area of its mean

**Divergence**   Evolution of closely related species in different directions ( = **divergent evolution**)

**Division**   An alternative name for a phylum, used mainly by botanists and microbiologists

**Dominant**   A character as fully developed when the alleles determining it are heterozygous as when they are homozygous (see **recessive**)

**Duplication**   The reduplication of one or more segments of a chromosome resulting from abnormal cell division ( = **repetition**); it results in an increase in the number of genes in the nucleus

**Ecological genetics**   The ecological study of adaptations in wild populations of plants and animals through the application of genetics and appropriate mathematical analysis

**Electrophoresis**   A technique of separating mixtures by the migration of electrically charged particles in solution towards oppositely charged electrodes under an electric field

**Enantiomers**   Molecules that can exist in two forms, the atomic structure of one being the mirror image of the other; they are characterised by the direction in which they rotate the plane of polarised light

**Endemic species**   Species whose distribution is confined to a particular area

**Environment**   The conditions in which living organisms exist; they can be divided into the physical (abiotic) environment (e.g. temperature) and the biotic environment deriving from the interactions of the organisms themselves (e.g. food supply)

**Era**   The largest division of geological time; the three eras since the Precambrian are the Palaeozoic, Mesozoic, and Cenozoic

**Erosion**   The wearing away of rock by forces such as rain and wind

**Eukaryotes**   A grouping of the four higher kingdoms of organisms on the basis of the type of cells they possess (see **prokaryotes**)

**Evolution**   Change in the characteristics of organisms occurring in successive generations in response to the environment in which they live

**Exon**   A sequence of nucleotides coding for a particular protein

**Extinction**   The disappearance of a species or other group of organisms from part or all of its previous geographical range

**Family**   A taxon often inserted between the order and genus; in animals the ending is **-idae** and in plants **-aceae**

**Fertility**   The number of progeny produced by a well-adapted individual relative to those of a less well-adapted one

**Fitness**   The adaptedness of an organism judged in terms of its fertility

**Fossil**   The dead remains of plants and animals preserved in rock, amber, or peat

**Founder principle**   The establishment of a new colony of a species from a few migrants carrying a gene frequency different from the average of the parent population

**Fragmentation**   Mutation resulting from an alteration in the structure of chromosomes

**Fundamentalist**   Originally, anyone adhering to the fourteen fundamental doctrines laid down at the 1878 Niagara Bible Conference. Now applied to those believing in the absolute truth of the Bible, particularly the accounts of the Creation in the Book of Genesis

**Gamete**   A haploid cell capable of fusing with another haploid cell to form a diploid zygote (see **zygote**)

**Gap theory**   A religious theory which accounts for fossils as the remains of previous life that existed on Earth before the Creation as described in the Book of Genesis

**Genes**   Units of heredity; they produce a given set of characters in any particular environment

**Gene pool**   The total of the genes present within a breeding population at a particular time

**Genetic drift**   Fluctuation of gene frequency in a small population due to chance; it is one of the factors that can upset the Hardy-Weinberg equilibrium

**Genetic polymorphism**   The occurrence together of two or more forms of the same species, the rarest existing at a frequency above that of recurrent mutation

**Genotype**   The gene complement carried by an organism (see **phenotype**)

**Geographical distribution**   The occurrence of populations judged in relation to the zones that the organisms occupy

**Germ cells**   Cells in animals that give rise to gametes (see **somatic cells**)

**Germline theory**   A theory of immunology which states that all antibody specificities are encoded in the genes of an animal and therefore inherited *en bloc*

**Gradualism**   Gradual evolutionary change – as envisaged by Darwin

**Haploid**   The condition in a cell or organism where only one representative from each pair of homologous chromosomes is present

**Hardy–Weinberg equilibrium**   The theoretical situation in a large randomly mating population in which the proportion of dominant and recessive genes remains constant from one generation to the next

**Heterosis**   Improved phenotypic effects (e.g. more rapid growth) among plant cultivars that are heterozygous at a large number of gene sites (=**hybrid vigour**)

**Heterozygous advantage**   A situation where a gene is disadvantageous as a homozygote but beneficial as a heterozygote

**Homology**   Structures occurring in different species are said to exhibit homology if they have the same basic pattern and are therefore assumed to have a common ancestry (see **analogy**)

**Hybrid vigour**   See **heterosis**

**Indicator species**   Plants characteristic of particular ecological environments

**Interglacial**   The period between two glacials (ice ages)

**Intermittent drift**   Random gene assortment occurring when a population reaches its lowest density as a result of a fluctuation in numbers

**Interspecific competition**   Competition occurring between different species

**Intraspecific competition**   Competition between individuals of the same species

**Introns**   Inert sections of DNA that have no genetic effects and are lost at the formation of messenger RNA

**Inversion**   Chromosome fragmentation involving a change in the sequence of genes

**Isolation**   The prevention of gene flow within part of a population; a powerful influence on the development of new varieties and species

**Isotopes**   Atoms of the same element but of different atomic weights

**Kingdom**   The largest classificatory group

**Kinship selection**   The sum of altruistic acts quantified according to their genetic significance

**Lamarckian evolution**   A mechanism of evolution propounded by the Frenchman Lamarck based on the inheritance of the effects of use and disuse; it assumes an environmental control of the process of mutation

**Linkage**   The tendency of certain genes to remain together instead of assorting independently, since they are carried on the same chromosome

**Macroevolution**   Evolution occurring at the species level and above (see **microevolution**)

**Maternal inheritance**   A mechanism of inheritance in which the effects of the cytoplasm of the egg override those of the nucleus

**Meiosis**   The process of cell division in which the number of chromosomes in the daughter cells $(N)$ is half that of the parent $(2N)$

**Melanism**   The deposition of the pigment melanin resulting in black or dark-brown individuals, mostly in insects

**Microevolution**   Evolution below the species level (see **macroevolution**)

**Mitochondria**   Organelles in the cells of plants and animals associated with aerobic respiration; they also contain DNA and have a hereditary function

**Mitosis**   The usual process of somatic cell division in which the chromosome number remains constant $(2N)$

**Multiple alleles**   A series of genes occurring at the same place on a chromosome having arisen by mutation (see **polygenic inheritance**)

**Mutagen**   A substance that increases the mutation rate

**Mutation**   A change in one or more of the bases in DNA resulting in the formation of an abnormal protein. Mutations are only inherited if they occur in the germ cells destined to give rise to gametes

**Natural classification**   A taxonomic system based on homologous resemblances

**Natural selection**   A process in which those organisms best adapted to their environment tend to survive and breed at the expense of the less well adpated

**Neo-Darwinism**   The theory of evolution derived from the ideas of Darwin

**Neoteny**   Precocious sexual development accompanied by a relative reduction in the rate of bodily growth; almost certainly under genetic control

**Non-disjunction**   The failure of homologous chromosomes to move to different poles at anaphase of the first meiotic division. Two of the four gametes formed are therefore without one chromosome ($N - 1$)

**Non-nuclear inheritance**   See **cytoplasmic inheritance**

**Normal distribution**   A quantitative variable whose frequency conforms to a bell-shaped curve

**Nucleotide**   A unit of DNA consisting of a base, sugar, and phosphate

**Numerical taxonomy**   A taxonomic system in which numerical values are assigned to different degrees of resemblance between organisms

**Ontogeny**   The development of an organism from egg to adult

**Order**   A group of similar genera

**Orthogenesis**   The tendency for related groups of organisms to evolve in the same direction

**Particulate inheritance**   The type of inheritance (postulated by Mendel) involving discrete genetic units (genes) which do not blend with one another

**Phenotype**   The characteristics of an organism resulting from the reaction of a given genotype with a particular environment (see **genotype**)

**Phylogenetic tree**   The relationship of a group of organisms expressed as a branching diagram based on overall similarities, some of which are homologous

**Phylum**   A group of similar classes

**Pleiotropic effects**   Multiple effects controlled by a single pair of alleles

**Polygenic inheritance**   The inheritance of a character due to the action of genes carried on different chromosomes which therefore assort independently (see **multiple alleles**)

**Polymer**   A large molecule formed by the joining together of two or more similar molecules

**Polymorphism**   See **genetic polymorphism**

**Polytypic species**   A species containing a number of different variants

**Polyploidy**   Situations in which more than two members of each gene pair are present in the cells of an organism

**Polysomy**   Abnormality in diploid ($2N$) cells with one chromosome occurring once only or more than twice

**Position effect**   A situation in which the phenotypic expression of a pair of alleles is governed by their position relative to other alleles (e.g. in chromosome repetition)

**Postglacial**   The period after an ice age (glacial)

**Progressionism**   The idea (propounded by Cuvier) that after each catastrophe the organisms that followed tended to be more advanced than those before

**Prokaryotes**   Cells with a primitive type of structure characteristic of the Monera (see **eukaryotes**)

**Protocells**   Associations of organic molecules containing compounds essential for life

**Psychosocial evolution**   Evolution resulting from the accumulation of knowledge and the passing on of experience

**Punctuational evolution**   A pattern of evolution consisting of a series of discontinuous (punctuated) episodes (see **gradualism**)

**Radiation**   See **adaptive radiation**

**Recapitulation**   An idea ('Law') propounded by Haeckel that during the course of its development an animal ascends its ancestral tree

**Recessive**   A character expressed only when the alleles controlling it are homozygous (see **dominant**)

**Repetition**   See **duplication**

**Ribosome**   An organelle occurring in large numbers in the cytoplasm of cells and acting as a site for protein synthesis

**Sedimentary rocks**   Rocks formed in layers as a result of erosion; compare igneous rocks formed by volcanic action

**Selective agents**   Ecological factors (physical or biotic) responsible for natural selection

**Somatic cells**   All the bodily cells of animals other than those concerned in the formation of gametes (see **germ cells**)

**Species selection**   The natural selection of species as groups rather than of individuals

**Stabilising selection**   Selection of a quantitative variant such that the optimum corresponds to the mean, the two extremes being disadvantageous

**Supergene**   Two or more genes controlling different characters so closely linked that they act as a single switch mechanism governing alternative forms

**Symbiosis**   The close association between two or more different organisms to their mutual benefit

**Sympatric evolution**   Evolution of a group of organisms in the absence of any physical barrier isolating them from the parent stock (see **allopatric evolution**)

**Taxonomy**   The classification of organisms into groups (taxa)

**Transient polymorphism**   A situation in which a polymorphic species changes from one equilibrium to another (see **balanced polymorphism**)

**Translocation**   The transfer of a segment from one chromosome to another; it results in an increase in the number of genes in the nucleus

**Type specimen**   A specimen regarded by taxonomists as typical of a species

**Uniformity**   A principle enunciated by Lyell that the same forces are involved in the formation of the Earth's crust today as in the past; everywhere they are of the same kind but they may act at differing rates

**Vestigial structures**   Organs which were once functional but which have since been reduced to vestiges in the course of evolution

**Zygote**   A diploid cell resulting from the fusion of two haploid gametes (see **gamete**)

# Index